Developments in Petroleum Science
Volume 60

Geophysics for Petroleum Engineers

Developments in Petroleum Science

Volume 60

Geophysics for Petroleum Engineers

Fred Aminzadeh
Professor, University of Southern California, California

Shivaji N. Dasgupta
President, Reservoir Consultants Inc., Houston, TX
Previously Saudi Aramco
Dhahran Saudi Arabia

ELSEVIER AMSTERDAM · BOSTON · HEIDELBERG · LONDON · NEW YORK · OXFORD
PARIS · SAN DIEGO · SAN FRANCISCO · SINGAPORE · SYDNEY · TOKYO

Elsevier
Radarweg 29, PO Box 211, 1000 AE Amsterdam, The Netherlands
The Boulevard, Langford Lane, Kidlington, Oxford OX5 1GB, UK

First edition 2013

Notice
No responsibility is assumed by the publisher for any injury and/or damage to persons or
property as a matter of products liability, negligence or otherwise, or from any use or operation
of any methods, products, instructions or ideas contained in the material herein. Because of
rapid advances in the medical sciences, in particular, independent verification of diagnoses and
drug dosages should be made

British Library Cataloguing in Publication Data
A catalogue record for this book is available from the British Library

Library of Congress Cataloging-in-Publication Data
A catalog record for this book is available from the Library of Congress

ISBN: 978-0-444-50662-7
ISSN: 0376-7361

For information on all Elsevier publications
visit our web site at store.elsevier.com

Printed and bound by CPI Group (UK) Ltd, Croydon, CR0 4YY

Working together
to grow libraries in
developing countries

www.elsevier.com • www.bookaid.org

Contents

As the world's demand for petroleum continues to increase, there is a continual need for reservoir characterization for enhanced oil recovery in order to meet these demands. As time passes, the number of new discoveries of giant conventional oil fields is unlikely to meet these demands; however, there is still much oil in existing fields since recovery rates are modest. Most future oil production will come from presently existing fields. The science of reservoir characterization will undoubtedly involve the integration of information from the disciplines of geology, geophysics and reservoir engineering. There is a growing need for professionals in geoscience and engineering to educate each other in a synergistic integrated fashion. For these reasons, the book "Geophysics for petroleum engineers" represents a significant contribution to the cross-disciplinary education needed in the petroleum industry. Chapter 1 gives a comprehensive introduction to reservoir geophysics and its applications.

Appropriately, in Chapter 2, the book initiates reservoir discussions with a review of petroleum geology fundamentals including the creation of petroleum itself from the burial of organic matter and the sedimentation processes. A review of clastic and carbonate reservoir rocks is given, and the conditions for petroleum accumulation are outlined.

In the search for oil and gas deposits, it is imperative that the petroleum geoscientist describes the Earth's interior. One of the primary tools for doing this, is the seismic reflection method. Chapter 3 describes the acquisition, processing and interpretation of seismic data for both land and marine environments. Key aspects of the seismic method include resolution and characterization of subsurface geology with recently developed methods using seismic attributes, spectral decomposition and coherency analysis. The properties of fluid saturated rocks are described by amplitude variation with offset (AVO). While the first seismic exploration for oil and gas traps involved single (vertical) component recording, Chapter 3 points out that multicomponent recording adds valuable information to our knowledge of different lithologies within the reservoir. As pointed out in this chapter, surface seismic methods are not the only geophysical tools. The physical properties can be described by geophysical surveys such as gravity, magnetics, electrical and electromagnetic surveys. Borehole geophysical methods such as vertical seismic profiling (VSPs) and cross-borehole seismic methods add high resolution seismic information about the reservoir.

Detailed information about formations and fluids in the vicinity of the borehole can be achieved through the use of different well logs as described in Chapter 4. Logs that describe lithology changes include gamma ray logs, photoelectrical logs and spontaneous potential (SP) logs. For example, such logs are very effective in discriminating sandstone versus shale in the subsurface. Logs sensitive to the porosity of reservoir rocks include bulk density, neutron, nuclear magnetic resonance (NMR) and acoustic (sonic) logs. Well logs sensitive to reservoir fluids include NMR and resistivity logs.

Well logs, borehole geophysical methods and the seismic method measure different volumes of the reservoir with different degrees of resolution. In some sense, reservoir characterization involves using these measurements on different scales and different volumes in order to produce an accurate model of the petroleum reservoir. Geostatistics, as described in Chapter 5, attempts to integrate all available reservoir data. While geostatistics includes the classical prediction methods such as kriging and co-kriging, recent unconventional methods such as fuzzy logic, neural networks and genetic algorithms attempt to optimize the geostatistical models derived from all data. Where geology can be reliably predicted from available data, geostatistics is a powerful tool for describing the reservoir.

Chapters 6 and 7 examine the reservoir in a static, and in a sense that is dynamically related to production of hydrocarbons. With production, there are changes in physical characteristics of the reservoir that can be monitored using repeated geophysical surveys such as time-lapse or 4D seismic surveys. Such monitoring of subsurface changes will allow for optimization of drilling locations and drilling schedules in enhanced oil recovery (EOR).

Seismic information has traditionally been gathered from seismic sources such as dynamite, vibroseis, and marine air guns. However, seismic data can be obtained while drilling a well. Chapter 8 describes seismic measurements while drilling that can help to direct the drill bit and which can predict the presence of hazardous overpressure zones.

Finally for most of the 20th century, much oil and gas was produced from rocks with high porosity and permeability that contained low-viscosity fluids. It was predicted by M. King Hubbert that oil supplies from these conventional reservoirs would experience peak production early in the 21st century. However, these predictions of peak oil have proven to be inaccurate due to discovery and production of unconventional reservoirs. Oil and gas are now produced from tight formations of shale and carbonate through the process of hydraulic fracturing. Major oil production also comes from bituminous sands that contain high viscosity heavy oil. In both the case of tight oil and heavy oil production, time-lapse seismology plays a key role in reservoir characterization. Chapter 9 describes the important role of geophysics in EOR from these unconventional reservoirs.

Given the interest and importance of geophysics in characterizing conventional and unconventional petroleum reservoirs, it is anticipated that this book on "geophysics for engineers" will be a valuable source of information for those utilizing geology, geophysics and reservoir engineering in enhanced oil recovery.

Larry Lines
Department of Geoscience, University of Calgary

Geophysical techniques have proven to be effective tools for oil and gas exploration over the last hundred years. Although geophysical tools were originally developed for petroleum exploration, they are now increasingly used in reservoir management for optimum production from a reservoir. With the recent increases in the need for reservoir monitoring, enhanced oil recovery, and horizontal drilling, geophysical tools are in more demand. As a result, petroleum engineers are finding it necessary to become more proficient in the use of geophysical methods and to better understand their applications and limitations. We explore the complementary features of geophysical techniques in better understanding, characterizing, producing, and monitoring of the reservoirs. The objective of this book is to introduce the engineers to geophysical methods so that they can communicate their needs to geophysicists and realize the full benefits of geophysical measurements.

There are many books that address different aspects of geophysical technologies and seismic methods. Most of those books are rather specialized and primarily geared toward geophysicists. They focus on theoretical discussions on principles of wave propagation, acoustic or elastic wave equations in the subsurface, and describe different approaches for geophysical inversion. There are also specialized books on seismic processing, migration, and imaging, as well as geophysical data acquisition or interpretation of geophysical data.

This book fills the void, by emphasizing the aspect of geophysics that matters the most to the petroleum engineers. Following a general overview of the fundamentals of geophysics, we focus on specific applications of different seismic and other geophysical methods in the problems of petroleum engineering. We discuss both the static and dynamic aspects of reservoir characterization. We demonstrate how time lapse seismic data integrated with other geophysical and well data can highlight changes in the reservoir fluid and reservoir pressure. We also address the drilling applications of geophysics including "geo-steering." Furthermore, we look at the applications of geophysics in the development of shale and tight sand reservoirs, including application of microseismic data for monitoring of hydraulic fracturing.

For completeness, we also include three complementary chapters which we consider important for engineers to get the full benefits of this book. One chapter covers the fundamentals of petroleum geology. Another chapter gives an overview of petrophysics and well log analysis, which has been considered as part of geophysics by many. Yet another chapter provides an overview of geostatistical methods. Aside from the conventional statistical methods such as kriging and co-kriging for reservoir characterization, we discuss unconventional statistics or "Soft Computing" methods such as neural networks, fuzzy logic, genetic algorithms, as well as the hybrid methods.

This book is based on material from many sources including various Society of Exploration Geophysicists, American Association of Petroleum Geology, European Association of Geoscientists & Engineers, and Society of Petroleum Engineers publications. Many other sources including material from websites of different oil and gas companies, academic institutions, and geophysical contractors are used. We have attempted to refer to the sources of those materials to the best of our abilities and we acknowledge possible omissions and are grateful nevertheless.

We wish to thank Prof. Lawrence Lines of the University of Calgary for writing an excellent foreword to the book. We also thank Prof. Jamie Rector and Dr. Don Hill for their important contributions to the book. In addition, we would like to acknowledge many reviewers of different chapters including Steve Hill, Kelly Rose, Wenlong Xu, and Kurt Strack. Finally, we are thankful for the encouragement and patience of Elsevier editorial staff and the series editor, John Cubitt.

Fred Aminzadeh
Professor, University of Southern California, California

Shivaji N. Dasgupta
Houston, Texas
December 2013

Series Editor's Preface

It is a pleasure for me to introduce this volume on *Geophysics for Petroleum Engineers*, written by Professor F. Aminzadeh and Dr. S.N. Dasgupta. Until the last 30 years, the exploration for oil and gas was conducted primarily by geologists and geophysicists who worked together to locate suitable sites for the drilling of wells to test hydrocarbon-bearing structures. Once potentially commercial quantities of oil or gas were discovered, the fields were then turned over to engineers who designed, constructed, and applied systems to develop and exploit the hydrocarbons. However, these two approaches were often conducted in almost total technical isolation, leading to frequent suboptimal development of fields.

Now the more normal strategy is to establish asset teams consisting of a mixture of technical disciplines from geologists through to engineers who work together to optimally develop fields. This leads to greater understanding and appreciation of the contributions each discipline makes to a commercial development. However, the language, terminology, concepts, and technology used by these disciplines are complicated and still lead occasionally to confusion or misunderstanding. So there remains a strong need to maintain and improve lines of communication among and between modern reservoir, drilling, and production engineers and geophysicists, geologists, and petrophysicists. This volume, produced by experts in the geosciences and engineering within industry and academia, has been designed and written with alleviating this communication gap firmly in mind.

This preface also provides an opportunity to inform readers that from this volume, the *Developments in Petroleum Science* series will now incorporate the *Handbook of Petroleum Exploration and Production* series going forward. As currently Series Editor of the *Handbook of Petroleum Exploration and Production*, I will continue as Series Editor of the *Developments in Petroleum Science* series and hope that readers will enjoy the combined series.

John Cubitt
Holt, Wales

CHAPTER 1   INTRODUCTION

This book is a treatise on geoscience disciplines with a focus on geophysical application to petroleum engineering. While the book's focus is on geophysical applications, the chapters delve into other related disciplines that participate in the process. Petroleum engineers require some working knowledge of geology and geophysics during the different stages of development of oil and gas fields. Reservoir, Drilling, and Production Engineers must be able to understand the information provided by Geophysicists, Geologists, and Petrophysicists for properly utilizing it. This is particularly important because of the multidisciplinary nature of the challenges faced in oil and gas exploration and production. The need to integrate the data and the disciplines is important.

This process begins with exploration, discovery and appraisal drilling through reservoir development, production and enhanced recovery, as well as its eventual depletion and abandonment. The team efforts by geoscientists and engineers focus on maximizing economic recovery of hydrocarbons throughout the life of a field. Integration of geophysical data with geologic data, and engineering measurements improves our understanding of the reservoir, reduces uncertainties, and mitigates the risk. The improved knowledge of the reservoir impacts the life of the field, its economics, and the ultimate recoverable volume of oil and gas from the field resource base. A detailed understanding of the physical behavior of oil, water, and gas within porous rocks at reservoir pressure and temperature and their impact on the characteristics of the geophysical measurements are ascertained.

CHAPTER 2   PETROLEUM GEOLOGY

Understanding the basics of Petroleum Geology is critical for Petroleum Engineers. The chapter begins with the formation of organic matter and the origin of petroleum from burial of organic matter and the sedimentation processes. Petroleum systems are then discussed, which comprise Source Rock, Burial Depth and Temperature, Reservoir Rock, Migration Pathways, Reservoir Seals, and Traps. Different types of petroleum traps such as anticlinal, fault, salt-related, and stratigraphic traps and various types of reservoir rocks such as clastic (sandstone and shale) and carbonate rocks are enumerated. The conditions for petroleum accumulation in the reservoir are outlined. The chapter concludes with the integration of geology, geophysics, and petrophysics, in connection with reservoir geometry, volume, and assessment of reserves.

CHAPTER 3   PETROLEUM GEOPHYSICS

Geophysical techniques apply the principles of physics for study of the earth. Geophysics is the study of physical responses of rocks under passive or active perturbation. Data from geophysical observations are interpreted to infer geology. Multiple geologic parameters are assessed with the same geophysical data. Geophysics measures changes of physical properties. The data interpretation has inherent ambiguity, that is, multiple interpretations. Data from geophysical tools provide coverage with spatially continuous high density of 10–25 m and vertical resolution of the order of 10–20 m. Well data like cores and well logs provide vertically high resolution of the order of 0.5 m or better at the well location; however, the distribution of wells is sparse and discontinuous. The detailed spatial coverage from geophysical data is calibrated with analysis of well logs, pressure tests, cores, geologic depositional knowledge, and other information from appraisal wells.

Geophysical data play an important role in the development of a gross reservoir model. The reservoir architecture (structure) and the reservoir properties are derived from the analysis and integration of data from various geoscience disciplines. The distributions of the reservoir and non-reservoir rock types and of the reservoir fluids determine the geometry of the model and influence the type of model to be used. Thus, the goal of geophysics is to contribute to the increment in spatial resolution for defining the building blocks of the reservoir. Geophysics contributes by either adding value or by preventing loss. The data interpretation is used for guiding business decisions. Geophysical data acquisition, processing, and interpretation are driven by established scientific principles.

The objective of geophysical techniques is to minimize risk and maximize value. Exploration risk changes throughout the life of a venture. Geophysics contributes to reservoir characterization, reservoir monitoring, and its management by adding maximum value to improving production plan and by minimizing risk (risk of dry hole, risk of blow out, risk of inefficient recovery process, among others). Geophysics is a risk reduction tool; it reduces exposure to loss.

For optimum application of geophysical data for petroleum engineers, integration of many disciplines is essential. Geophysicists calibrate the measured geophysical attributes with rock properties near the wellbores. They use well logs, core data, and borehole seismic information that are available in order to test the correlation of reservoir data with geophysical measurements. Other reservoir properties that can affect geophysical measurements are density, oil viscosity, stresses, and fractures. Detailed understanding of reservoir rock and fluid properties and their influence on production and injection efficiency is imperative for optimum asset management. As the primary production from a reservoir begins, the development requirement is to position new wells at optimal locations that would maximize hydrocarbon recovery. During secondary recovery and then enhanced recovery process, the engineer's objective is to maximize the volume of hydrocarbon contacted by injected fluids. This is to achieve maximum volumetric sweep efficiency for fluid production. To minimize cost and risk, engineers attempt to predict reservoir performance—for both planning and evaluation of hydrocarbon recovery projects. Reservoir description in terms of reservoir architecture, flow paths, and fluid-flow parameters is the key to reservoir engineering. Accurate prediction of reservoir production performance is predicated primarily on how well the reservoir heterogeneities are understood and have been modeled and applied for fluid-flow simulation.

Ambiguity in seismic interpretation– lateral changes in amplitude can be caused by changes in one or more properties and are therefore inherently ambiguous. Structural features apparent on seismic data could be due to local anomalies unrelated to the structure.

Geophysical methods use high-precision sensors (e.g., geophone, hydrophone, magnetometer, and gravity meter) that measure the physical properties on the surface, in oceans, in wells, and from air. Rather than the overall magnitudes of these properties, the small differences in physical properties that exist among various rock bodies are what we need. These differences in physical properties must be measured accurately. Accuracy of measurements and their analysis rely heavily on the technological development. Geophysical tools are deployed from ground surface, at sea, in boreholes, and in air. There are also measurements from satellites.

CHAPTER 4   PETROPHYSICS

Petrophysical analysis of well logs and core provides information about formation rocks and fluids in the borehole. Various types of well logs measure different properties in the well. Logs that describe lithology changes include gamma ray logs, photoelectrical logs, and spontaneous potential (SP) logs. Such logs are very effective in discriminating sandstone versus shale in the subsurface. Logs sensitive to the porosity of reservoir rocks include bulk density, neutron, nuclear magnetic resonance (NMR), and acoustic (sonic) logs. Well logs sensitive to reservoir fluids include NMR and resistivity logs.

Analysis of the data determines the volume of hydrocarbons present in a reservoir and its potential to flow through the reservoir rock into the wellbore. This helps us to understand and optimize the producibility of a reservoir. When oil and gas wells are drilled, physical property measurements are taken from specialized geophysical instrument packages: either attached as drill collars, behind the drill bit Measurements While Drilling (MWD) or Logging While Drilling (LWD), or suspended on wireline cables (Wireline Logs) after the drill pipe has been removed from the borehole. Initially, these measurements were designed to provide detailed stratigraphic and structural correlation of geologic horizons between wells. In time, however, the measurements, themselves, and their application have much more complex, to the point that the future of wells and fields hinge on the interpretation of these measurements.

We cannot measure Porosity and saturation directly; we measure formation electrical galvanic or induction Resistivity (*R*), mud filtrate/connate water salinity contrast (Spontaneous Potential, or *SP*), formation radioactivity (Gamma Ray, *GR*), inverse acoustic velocity (Interval Transit Time or Δ_t), formation electron density Density Log (*RhoB*) and Photoelectric Effect (*PEF* or *Z*), formation hydrogen ion density (Neutron Log, *HI*), and Nuclear Magnetic Resonance (NMR Log).

CHAPTER 5   GEOSTATISTICS

A reservoir is intrinsically deterministic. In reservoir description process, we are dealing with limited and incomplete data. We are constantly trying to extrapolate information from sparse measurements (e.g., limited well data and core data on the one hand and large volumes of seismic data with limited spatial resolution on the other). We resort to statistical methods to accomplish this. Traditional statistical methods for both spatial and temporal extrapolation have been used in E&P for several decades. Among conventional statistical methods used are Matrix Plot, Correlation, Regression, Principal Component Analysis, Variogram, Kriging, and Clustering. One of the main uses of statistics has been for reservoir characterization through integrating information and data from various sources with varying degrees of uncertainty, such as log and seismic data. Other applications include establishing relationships between measurements and reservoir properties; and between reserve estimation and oil field economics with the associated risk factors. Stochastic techniques are applied to deterministic reservoirs because of

1. incomplete information about reservoir on all scales,
2. complex spatial deposition of facies,
3. variability of rock properties,
4. unknown relationships between properties,
5. the relative abundance of singular pieces of information from wells, and
6. convenience and speed.

Over the last two decades, a new brand of statistical methods, referred to as soft computing (SC), have found their way into many practical applications including the petroleum arena. Where conventional statistical means are deemed inadequate to tackle practical problems, we can employ nontraditional SC methods such as artificial neural networks, fuzzy logic, and genetic algorithms.

CHAPTER 6   RESERVOIR CHARACTERIZATION

Optimal reservoir development depends on the insight into the reservoir architecture. To achieve accuracy and to ensure that all the information available at any given time is incorporated in the reservoir model, reservoir characterization must be dynamic. The main objective of reservoir characterization is to transform the available seismic, log, geologic, production, and other data into reservoir properties. The reservoir properties include reservoir thickness, number of reservoir units, porosity, permeability, pressure distribution, fracture distribution (in the case of unconventional reservoirs), and fluid saturation (oil, gas, and water).

Reservoir engineers strive to recover maximum hydrocarbons from the reservoir. In order to achieve this, they need to understand the rock and fluid properties and their distribution in the reservoir system. The heterogeneities in the reservoir need to be characterized with some order of accuracy. Reservoir characterization process uses measurements from well logs, borehole geophysical instruments, and surface techniques like 3D seismic that measure different volumes at different scales. The process produces an accurate model of the petroleum reservoir in three dimensions. As new petrophysical, seismic, and production data become available, the reservoir model is updated to account for the changes in the reservoir. Both static reservoir properties, such as porosity, permeability, and facies type, and dynamic reservoir properties, such as pressure, fluid saturation, and temperature, need to be updated as more field data become available.

CHAPTER 7   RESERVOIR MONITORING

Monitoring of production-induced changes is crucial to sustain, optimize, and improve production levels and recoverable reserves. Increasing production efficiency and monitoring water/steam/CO_2 floods are key issues that are addressed with borehole and surface technologies and measurements. At the same time, linking the information to 3D surface and borehole seismic data requires extrapolation to the inter-well space. The goal of reservoir monitoring is to use all the available data to create a model for the reservoir with as accurate estimates of the reservoir properties as possible. Accurate prediction of reservoir performance relies on the proper definition of the frame of the reservoir, which is the rock matrix with empty pores. Reservoir characterization determines hydrocarbon distribution and the pathways or barriers impeding flow toward producer wells. As we produce from the reservoirs, new data become available. These include the production data, updated decline curves, and possibly new seismic data. Creating an updated reservoir model or "dynamic model" is an important step to better understand any important changes in the reservoir characteristic. This information is crucial to do a better job in reservoir management and optimize production. It is also important when we need to make certain interventions such as enhanced oil recovery (to increase permeability) or artificial lift (to increase pressure). It is also important to get updated information about the reservoir properties when we need to do an infill drilling. In short, we need to do an effective reservoir monitoring and surveillance during the producing life of a field, and mapping of oil–water and gas–oil interfaces is necessary for understanding the fluid dynamics.

CHAPTER 8   GEOPHYSICS IN DRILLING

The process of drilling an oil or gas well requires knowledge of all geologic features that are expected to be encountered along the way—from the surface of the ground to the target reservoir. Seismic measurements while drilling that can help to direct the drill bit which can predict the presence of hazardous overpressure zones. Thus, in addition to steering the well so as to intersect hydrocarbon-bearing reservoirs, drilling and reservoir engineers must assure to a reasonable degree of confidence that the well drills successfully and safely to the target.

By providing a picture of the subsurface from the surface to the target, geophysical measurements help ensure a successful drilling program. This geophysical picture helps to

1. identify drilling hazards that may lead to an uncontrollable well;
2. describe construction hazards and predict what lies ahead of and around the drill bit; and
3. illuminate what exists above and below the wellbore in a horizontal or highly deviated well

CHAPTER 9   GEOPHYSICS FOR UNCONVENTIONAL RESOURCES

The chapter addresses geophysical methods that are specifically more relevant to the exploration of and production from unconventional reservoirs. While many of the techniques have common applications for both conventional and unconventional reservoirs, there are also some significant differences in focus. Much of the unconventional reservoirs are from shale formations. Characterizing fracture system in such reservoirs is of most importance, not only to identify the "sweet spots" for well placement but also for optimum drilling and production from such fractured reservoirs. Combining conventional and micro earthquake seismic data has proven to improve the characterization process.

Another important factor for drilling through shale reservoirs is the need for stimulation through hydraulic fracturing. Use of microseismic data for monitoring the frac process has gained prominence in recent years. Different types of designs to acquire such data and utilize them for multistage fracking process as well as their integration with the conventional seismic data have been developed. This chapter addresses these issues and highlights different geophysical techniques suitable for unconventional resources at different stages of their exploration and exploitation.

Introduction

1.1 PURPOSE OF THE BOOK

This volume focuses on the application of geophysics to petroleum engineering disciplines. The objective is to introduce petroleum engineers to application of geophysical methods so that they can better communicate their needs and appreciate the full benefits derived from the application of geophysics. We hope this book will help engineers understand the integration of geophysical, geological, and petrophysical concepts and their applications. Understanding of the reservoir rock and fluid properties and their influences on production and injection efficiency is imperative for optimum asset management. Geophysical data integrated with well data can address this requirement. The chapters define the fundamentals of geophysical techniques, their physical basis, and their applications. They also describe the limitations of geophysical tools and the potential pitfalls in their misuse. Many real life examples illustrate the integration of geophysical data with other data types for predicting and describing reservoir rock and fluid properties.

Developments in Petroleum Science, Vol. 60. http://dx.doi.org/10.1016/B978-0-444-50662-7.00001-9

We emphasize that we have attempted to cover the most important topics that we believe will be of value to practicing engineers and petroleum engineering students who want to advance their understanding of geophysical technologies. To accomplish this task, certain compromises have had to be made. We provide, for instance, very little coverage to seismic data processing, which is important but may not be as crucial for a working engineer. Yilmaz (2001) is an excellent book addressing seismic processing and imaging issues. Lines and Newrick (2004) and Liner (2004) also cover processing and other geophysical concepts in depth. Many theoretical details on the wave equation the basis of most geophysical techniques, are also treated rigorously in these and many other books.

Instead of dwelling on technical details with theories and equations, we have used a large number of examples with figures from various sources and case histories to introduce different subjects. The idea is to convey useful information and provide examples of real life applications for those who may not want to get into in-depth studies on a given subject. The references, case histories, and examples we have included make no claim to be the most recent or most important ones but they are those we happen to know about. Given the wide area we have had to cover, we acknowledge possible omission of many important ones.

1.2 GEOSCIENCE DISCIPLINES

Geology is an observational science. It involves the study of the earth by direct measurements of rock properties, either from surface exposures (outcrops) or from boreholes, tunnels, and mines. Geological techniques allow deduction of the earth's structure, rock texture, composition, and history by the analysis of these observations.

Geophysics, on the other hand, applies the principles of physics to the study of the earth, for deducing physical features of the earth's surface and its internal structure. Geophysics involves the study of those parts of the earth hidden from direct view, by measuring their physical properties, with appropriate instruments on or above the surface of the earth, remotely from the measurement targets. Some geophysical tools measure physical responses of the ambient fields of rocks in a passive mode, for example, gravitational, magnetic, and radioactive. Other geophysical tools such as seismic, gravitational, electrical, and electromagnetic methods rely on either some active source of energy that transmits through the subsurface rocks or passive sources of energy (such as movement of the earth) or fracturing caused by stress, in the case of microearthquake data. The signal from the source of energy (either passive or active) is altered by the properties of rocks and this response is measured.

Virtually all of what we know about the earth below the limited depths to which boreholes, tunnels, and mines have penetrated, has been derived from geophysical observations. The properties of the solid inner core, the liquid outer core, the lower mantle and upper mantle, and the crust have all been deduced from the propagation of seismic waves from earthquakes.

Petrophysics deals with the physical and chemical properties of the earth's rocks and fluids. Petrophysicists provide the physical parameters upon which geophysical inversions are based and the detailed reservoir volumetric and flow properties upon which petroleum reserves are based. While geophysicists deal with indirect measurements on very large scales, which infer gross (heterogeneous) lithological units and structures, petrophysicists work on a much more detailed scale, looking at the various heterogeneities that geophysicists homogenize into gross structural units.

1.2.1   Geosciences in Petroleum Engineering

Petroleum engineers require some working knowledge of geology and geophysics during the different stages of development of oil and gas fields. This process begins with exploration, discovery, and appraisal drilling through reservoir development, production, and enhanced recovery, as well as the field's eventual depletion and abandonment. Team efforts by geoscientists and engineers focus on maximizing economic recovery of hydrocarbons throughout the life of a field. Integration of geophysical data with geologic data and engineering measurements improves our understanding of the reservoir, reduces uncertainties, and mitigates the risks. The improved knowledge of the reservoir impacts the life of the field, its economics, and the ultimate recoverable (EUR) volume of oil and gas from the field resource base. Some understanding of the physical behavior of oil, water, and gas within porous rock at reservoir pressure and temperature and their impact on the characteristics of the geophysical measurements are ascertained.

Reservoir, drilling, and production engineers must be able to understand the information provided by geophysicists, geologists, and petrophysicists so as to be able to properly utilize it. This is particularly important because of the multidisciplinary nature of the challenges faced in oil and gas exploration and production. The need to integrate the data and the disciplines is important. As engineers become more familiar with geophysical techniques, there will be expansion in applications of geophysical techniques in reservoir engineering practices. Geophysical tools are continuously evolving in order to address the present requirements and to be prepared for future challenges.

During a reservoir's life cycle from discovery and development to production and field maturation, the needs for reservoir description change continuously. Table 1.1 shows the reservoir analysis by integration during the various stages in the life cycle of a field.

1.3   GEOENGINEERING CONCEPT

Integrated asset management and encouraging geoscientists and engineers to work closely with each other have gained popularity in recent years. Aminzadeh (1996) introduced the concept of geoengineering as the wave of the future. He indicated, "As we approach the next millennium and as our

TABLE 1.1 Reservoir Analysis by Integration of Techniques During Field Production Life

Life Cycle of Field	Geophysical Techniques	Well Measurements Cores, Well Logs	Subsurface Modeling	Data Integration and Inversion
Discovery, appraisal	Structure Faults and fracture characterization Reservoir architecture Lithology Porosity distribution Hydrocarbon indicators	Reservoir facies Porosity, permeability Stratigraphy Hydrocarbon	Reservoir boundaries Faults, fracture mapping Oil–water contact	3D model of reservoir layers Reservoir connectivity mapping
Field development process	Reservoir architecture Fault sealing Lithology Porosity distribution Hydrocarbon indicators	Reservoir architecture details Fluid flow layers Fluid saturation distribution	Correlation of reservoir layers over field area	3D model of reservoir layers Flow layers Reservoir simulation
Production cycle and field maturation	Reservoir monitoring Time-lapse seismic, controlled source electromagnetic (CSEM)	Reservoir layering geometry Compartmentalization in reservoir	Fluid saturation distribution	Fluid saturation changes Sweep efficiency in reservoir

problems become too complex to rely only on one discipline to solve them more effectively, multidisciplinary approaches in the petroleum industry become more of a necessity than professional curiosity. We will be forced to bring down the walls we have built around classical disciplines such as petroleum engineering, geology, geophysics and geochemistry, or at the very least make them more permeable. Our data, methodologies, and approaches to tackle problems will have to cut across various disciplines. As a result, today's 'integration,' which is based on integration of results, will have to give way to a new form of integration, that is, integration of disciplines."

The geoengineering idea was picked up by many others, most notably Corbett (1997) where he maintained, "There is an ongoing debate (Aminzadeh, 1996) concerning the emergence of a new petroleum discipline – Geoengineering. This has been put forward as a solution to the problems facing the petroleum industry in integrating the disciplines. This contribution to the debate addresses the response of an academic institution to the challenge." Indeed, Heriot-Watt University in Edinburgh, UK, established the first Geoengineering program, under Dr. Patrick Corbett. Since then, many other universities have adopted the concept in one form or another. Various operating oil companies have followed the same trend and established multidisciplinary asset teams in their organizations.

1.3.1   Petroleum Geophysics

Geophysics is the study of physical responses of rocks under passive or active perturbation. Data from geophysical observations are interpreted to infer geology. Multiple geologic parameters are assessed with the same geophysical data. Geophysics measures changes in physical properties. The interpretations or inferences made from geophysical data are, however, somewhat nonunique. Data interpretation has inherent ambiguity, that is, multiple interpretations can be made from the same data. In addition to signals, the data contains noise. These issues are addressed by increasing data redundancy or sampling the same subsurface multitudes of times and using signal enhancement techniques in processing. The data gathering is usually designed in a uniformly sampled grid. New data are being collected with increasingly finer sampling as computer and electronics technologies are enhanced and hardware is becoming more reliable and cost effective. An example of ambiguity in seismic interpretation—lateral changes in amplitude can be caused by changes in one or more properties and are therefore inherently imprecise. For example, structural features apparent on seismic data could be the result of local anomalies unrelated to the structure or some of the traditional hydrocarbon indicators may be erroneous. Nevertheless, geophysics offers the best hope of obtaining useful data with a wide lateral and vertical coverage.

Geophysical techniques apply the principles of physics to the study of the earth. Data from geophysical tools provide coverage with a spatially continuous high sampling density of 10–25 m and a vertical resolution of the order of 10–20 m. Well data such as cores and well logs provide a vertically high resolution of the order of 0.5 m or better at the well location; however, the distribution of wells is sparse and discontinuous. The detailed spatial coverage from geophysical data is calibrated with analysis of well logs, pressure tests, cores, geologic depositional knowledge, and other information from appraisal wells.

Geophysical data play an important role in developing a gross reservoir model. The reservoir architecture (structure) and the reservoir properties are derived from the analysis and integration of data from various geoscience disciplines. The distributions of the reservoir and nonreservoir rock types and of the reservoir fluids determine the geometry of the model and influence the type of model to be used. Thus, the goal of geophysics is to contribute to the increment in spatial resolution for defining the building blocks of the reservoir.

Most petroleum geophysical tools were originally developed for exploration. They are now, however, being increasingly applied for reservoir development, monitoring, and management, that is, for optimizing fluid production from reservoirs. Reservoir management is a continuous process from field development through enhanced recovery and, eventually, depletion.

The objective of geophysical techniques is to minimize risk and maximize value. Exploration risks change throughout the life of a venture. Geophysics contributes to reservoir characterization, reservoir monitoring, and its management by adding maximum value to improving the production plan and by minimizing the risks of dry hole, blow out, leaving too much oil behind in the pipe, not penetrating the most prolific part of the reservoir, and inefficient recovery, among others.

Geophysics is a risk reduction tool; it reduces exposure to loss. The techniques either add value (resource discovery and improved reservoir management) or prevent loss (drilling hazards or dry holes). The data interpretation is used for guiding business decisions. Geophysical data acquisition, processing, and interpretation are driven by established scientific principles.

1.3.2 Geophysical Tools and Techniques

Geophysical methods use high-precision sensors (e.g., geophone, hydrophone, magnetometer, gravity-meter) that measure the physical properties onshore and offshore, in wells and from air. Rather than the overall magnitudes of these properties, the small differences in physical properties that exist among various rock bodies are what we need for interpretation. These differences in physical properties must be measured accurately. Accuracy of measurements and their analysis rely heavily on the technological development. Geophysical tools are deployed from ground surface, at sea, in boreholes, and in air. There are also measurements from satellites:

- *Surface*: seismic reflection 2D, 3D, 4D magnetics, gravity, electromagnetics.
- *Borehole*: vertical seismic profiling (VSP), cross well seismic, cross well electromagnetics, microseismic, borehole gravimeter (BHGM), nuclear magnetic resonance (NMR).
- *Aerial*: gravity, magnetics, remote sensing, LIDAR imaging.

In Chapter 3, we provide some details on different aspects of geophysical data. For example, what do geophysical tools measure? How are the measurements of time (reflection arrival delay time), frequency, and seismic reflectance amplitude used to pinpoint the reservoir structure, depth, porosity, lithology, fluid saturation, and permeability? We also provide some answers to the following questions:

- *Resolution*: How big does the container have to be for geophysical tools to respond to or sense it?

- *Detectability*: How sharp does the boundary have to be vertically and horizontally? What is the minimum change in physical properties that we can detect? If the geophysical tools cannot deduce the change, it does not matter if it is there or not.
- *Vertical resolution*: Depends upon integration of wavelets with reflecting surfaces and with each other. Closest separation of two wavelets in given bandwidth, usually 1/4 wavelength of central frequency (tuning thickness).
- *Horizontal resolution*: Depends on sampling frequency and on correct positioning of reflectors; data processing migration focuses on dispersed energy, and collapses diffractions.
- *Limit of visibility*: Bed thickness below which we can no longer distinguish signals from noise because the reflecting surfaces are too close to each other.

The next question is: Once we collect and process the Seismic Data, what do we do with them? The following are some of the key steps to follow:

- *Interpretation*: Deduce earth model from geophysical data and geological information.
- *Modeling*: Calculate seismic response from borehole data—logs, cores, fluids.
- *Synthetic seismogram*: Use velocity and density well logs to calculate the theoretical seismic responses of geologic sequences. This is used for calibrating surface seismic measurements with subsurface geology.
- *Inversion*: Compute response of a possible geologic sequence from seismic measurements.

1.4 INTEGRATION OF DISCIPLINES

For optimum application of geophysical data for petroleum engineering, integration of many disciplines is essential. Geophysicists calibrate the measured geophysical attributes with rock properties measured near the well bores. They use well logs, core data, surface measurements, and borehole seismic and other geophysical information that is available to test the correlation of reservoir data with geophysical measurements. In the integration process of different data sets, with varying scales, uncertainty, resolution, disparate sampling and environment are used. Conventional and unconventional statistical techniques are applied for addressing the challenge of assimilating the data to provide the best estimate for geologic and reservoir models from such integration. Other reservoir properties that can affect geophysical measurements are density, oil viscosity, stresses, and fractures. A detailed understanding of reservoir rock and fluid properties and their influence on production and injection efficiency is imperative for optimum asset management.

As the primary production from a reservoir begins, the development requirements are to position new production and injection wells at optimal locations that would optimize hydrocarbon recovery. During the secondary recovery and the enhanced recovery processes, the engineer's objective is to maximize the volume of hydrocarbon contacted by injected fluids. This is to achieve maximum volumetric sweep efficiency for fluid production. To minimize cost and risk, engineers attempt to predict reservoir performance—for both planning and evaluation of hydrocarbon recovery projects. Reservoir description in terms of reservoir architecture, flow paths, flow layers, and fluid-flow parameters is the key to reservoir engineering. Accurate prediction of reservoir production performance is dependent primarily on how well the reservoir heterogeneities are understood and on how they have been modeled and applied for fluid-flow simulation.

1.5 CONTINUOUS MEASUREMENTS IN INTELLIGENT FIELD

The new concept of i-fields (intelligent fields) or smart oil fields emphasizes placement of different sensors, including sensors that collect geophysical data and provide the effective integration of all the data collected. The real-time continuous data acquisition and data integration allow reservoir monitoring and, thus, the necessary guidance for various decisions that need to be made during the different phases of developing a field and producing from it. This is an important part of the next generation of oil fields for maximizing the EUR from the reservoir.

Figure 1.1 illustrates a typical i-field implementation: the system continuously monitors reservoir conditions, and the data are used in reservoir development decisions.

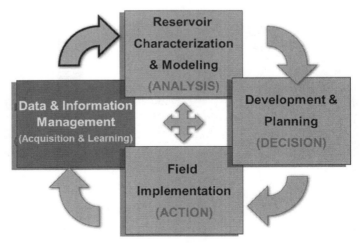

FIGURE 1.1 The closed loop from the data collection phase to the analysis, decision, and action phases.

While the book's focus is on geophysical applications, the various chapters delve into other related disciplines that participate in the process. The following is a synopsis of the related disciplines and applications that are included in this volume.

1.6   PETROLEUM GEOLOGY

Petroleum geology and its application to the reservoir process is critical for petroleum engineers. The geologist develops a viable model of the subsurface based on sparse observations in well logs, cores, and outcrops. The model needs to be consistent with geologic principles and based upon the model, exploration and/or development programs are planned. The chapter on petroleum geology begins by describing the formation of organic matter and the origin of petroleum. This is followed by a discussion on petroleum systems comprising source rock, burial depth and temperature, reservoir rock, migration pathways, reservoir seals, and traps. Petroleum systems occur in reservoirs within sedimentary basins in those areas of the world where subsidence of the earth's crust has allowed the accumulation of thick sequences of sedimentary rocks. Different types of petroleum traps such as anticlinal, fault, salt-related and stratigraphic traps and various types of reservoir rocks such as clastic (sandstone and shale), and carbonate rocks are enumerated. A seal rock keeps the oil entrapped in the reservoir and prevents it from migrating away. Understanding the geological and geomechanical nature of the seals is vital for successful exploration and reservoir development efforts. The discussion concludes with the integration of geology, geophysics, and petrophysics with respect to reservoir geometry, volume, and assessment of reserves.

1.7   FORMATION EVALUATION—PETROPHYSICS

Petrophysics data analysis determines the volume of hydrocarbons present in a reservoir and its potential to flow through the reservoir rock into the well bore. Well logs and core measurements are used in this analysis. This study helps us to understand and optimize the producibility of a reservoir. When oil and gas wells are drilled, physical property measurements are taken from specialized geophysical instrument packages, attached as drill collars behind the drill bit, such as Measurements While Drilling (MWD), Logging While Drilling (LWD), or suspended on wireline cables (Wireline Logs) after the drill pipe has been removed from the borehole. Initially, these measurements were designed to provide detailed stratigraphic and structural correlation of geologic horizons between wells. With time, however, the measurements themselves and their application have become much more complex, to the point that the future of wells and fields hinges on the interpretation of these measurements.

While we cannot measure porosity, and saturation directly, we can measure formation of electrical, galvanic, or induction resistivity (R), mud filtrate/connate water salinity contrast (spontaneous potential, or SP), formation of

TABLE 1.2 Formation Properties Measured by Common Open-Hole Formation Evaluation Logs and the Rock Properties Affecting Them

MWD/LWD or Wireline Measurement	Rock and Fluid Properties
Galvanic/induction resistivity log, R	Lithology, porosity, water saturation, and water salinity
Spontaneous potential log, SP	Water/mud filtrate salinity contrast
Gamma ray log, GR	Lithology, natural radioactivity
Acoustic/sonic log, Δ_t	Lithology, porosity, and gas/oil/water saturation
Density log, RHOB	Electron density, lithology, porosity, and gas/oil/water saturation
Photo-electric effect log, PEF or Z	Lithology
Neutron log or HI	Hydrogen ion density, lithology, porosity, and gas/oil/water saturation
Nuclear magnetic resonance	Proton density, porosity, pore lining materials, pore size (permeability)

radioactivity (gamma ray, GR), inverse acoustic velocity (interval transit time, or Δ_t), formation of electron density density log ($RhoB$) and photoelectric effect (PEF or Z), formation of hydrogen ion density (neutron log, HI), and nuclear magnetic resonance (NMR log) (Table 1.2).

1.8 GEOSTATISTICS

Haldorsen and Damsleth (1990) explained that stochastic techniques are applied to deterministic reservoirs because of (1) incomplete information about reservoirs at all scales, (2) complex spatial deposition of rock facies, (3) variability of rock properties, (4) unknown relationships between properties, and (5) the relative abundance of singular pieces of information from wells, as well as for (6) convenience and speed. A reservoir is intrinsically deterministic.

In the reservoir description process, we deal with limited and incomplete data. We are constantly trying to extrapolate information from sparse measurements (e.g., limited well data and core data on the one hand and large volumes of seismic data with limited spatial resolution on the other). We resort to statistical methods to accomplish this. Traditional statistical methods for both spatial and temporal extrapolation have been used in E&P for several decades. Among the conventional statistical methods used are matrix plot, correlation, regression, principal component analysis, variogram, kriging, cokriging, and clustering. For example see Deutsch and Journel (1998). One of the main uses of statistics has been for reservoir characterization through integration of information and data with varying degrees of uncertainty such as log and seismic data from various sources. Other applications include establishing relationships between measurements and reservoir properties, reserve estimation, and oil field economics with the associated risk factors.

Over the last two decades, a new brand of statistical methods, referred to as soft computing (SC) has found its way into many practical applications including the petroleum arena. Where conventional statistical means are deemed inadequate to tackle practical problems, we can employ nontraditional SC methods such as artificial neural networks, fuzzy logic and genetic algorithms.

1.9   RESERVOIR CHARACTERIZATION

Reservoir engineers strive to optimize the recovery of hydrocarbons and maximize the estimated EUR from the reservoir. In order to achieve this, they need to understand the rock and fluid properties and their distribution in the reservoir system. The heterogeneities in the reservoir need to be characterized with some degree of accuracy. Optimal reservoir development depends on a greater insight into the reservoir architecture. In order to achieve the needed accuracy and to ensure that all the information available at any given time is incorporated in the reservoir model, reservoir characterization must be dynamic. As new petrophysical, seismic, and production data become available, the reservoir model is updated to account for the changes in the reservoir. Both static reservoir properties, such as porosity, permeability, and facies type, and dynamic reservoir properties, such as pressure, fluid saturation, and temperature, need to be updated as more field data become available. The dynamic reservoir properties are monitored as production proceeds. The static model and the dynamic measurements are used in the numerical reservoir simulation models.

As Robertson (1989) suggests, the main objective of reservoir characterization is to transform the available seismic, log, geological, production, and other data to reservoir properties. The reservoir properties include reservoir thickness, number of reservoir units, porosity, permeability, pressure distribution, fracture distribution in the case of unconventional reservoirs, and fluid saturation (oil, gas, water).

1.10   RESERVOIR MONITORING

Increasing production efficiency and monitoring of water/steam/CO_2 floods are key issues that can be addressed with borehole and surface technologies and measurements. Monitoring of production-induced changes is crucial to sustain, optimize, and improve production levels and recoverable reserves. At the same time, linking the information to 3D surface and borehole seismic data requires extrapolation to the inter-well space. The goal of reservoir monitoring is to use all the available data to create and update a model for the reservoir with an accurate estimate of the changes in reservoir properties with production and injection. Accurate prediction of reservoir performance relies on the proper definition of the frame of the reservoir, which is the rock matrix with empty pores. Reservoir characterization determines hydrocarbon distribution and the pathways or barriers impeding flow toward producer wells. As we produce from the reservoirs, new data becomes available. This includes the production data, updated decline curves, and possibly new seismic data. Creating an updated reservoir model or "dynamic model" is an important step toward a better understanding of any important changes in the reservoir characteristics. This information is crucial for doing a better job at reservoir management and for optimizing production. It is also important when we need to make certain interventions such as enhanced oil recovery (to increase permeability) or artificial lift (to increase pressure). It is also important to get updated information about the reservoir properties when we need to do an infill drilling. In short, we need to do an effective reservoir monitoring and surveillance during the producing life of a field and mapping of oil–water and gas–oil interfaces is necessary for understanding the fluid dynamics.

1.11   GEOPHYSICS IN DRILLING

The process of drilling an oil or gas well requires knowledge of all geologic features expected to be encountered along the way—from the surface of the ground to the target reservoir. Thus, in addition to steering the well so as to intersect hydrocarbon-bearing reservoirs, the reservoir engineer must ensure that the well drills successfully and safely to the target formations.

By providing a picture of the subsurface from the surface to the target, geophysical measurements help ensure a successful drilling program. This geophysical picture helps to:

1. Identify drilling hazards that may lead to an uncontrollable well;
2. Describe construction hazards; predict what lies ahead of and around the drill bit; and
3. Illuminate what exists above and below the well bore in a horizontal or highly deviated well.

1.12   GEOPHYSICS FOR UNCONVENTIONAL RESOURCES

Unconventional resources refer to a recent trend that has been very successful in the production of gas and oil from source rocks with extremely low permeabilities. These formations are now considered as unconventional gas and oil reservoirs. The exploitation of these resources applies new innovative geophysical methods that are specifically more relevant to the exploration of and production from unconventional reservoirs. While many of the techniques have common applications for both conventional and unconventional reservoirs, there are also some significant differences in focus. For example, since many of the unconventional reservoirs are from shale formations, characterizing the fracture system in such reservoirs is of utmost importance, not only to identify the "sweet spots" for well placement but also for optimum drilling and production from such fractured reservoirs. Combining conventional and microearthquake seismic data has proven to improve the characterization process.

Another important factor for drilling through shale reservoirs is the need for stimulation through hydraulic fracturing. The use of microseismic data for monitoring the frac process has gained prominence in recent years. Different types of design to acquire such data and utilize them for multistage fracking processes as well as their integration with the conventional seismic data have been developed. This chapter addresses these issues and highlights different geophysical techniques suitable for unconventional resources at different stages of their exploration and exploitation.

REFERENCES

Aminzadeh, F., 1996. Geo-engineer, the wave of the future. J. Petrol. Sci. Eng. 15 (1), vii–x.

Corbett, P.W.M., 1997. Geoengineers: A Subject of Debate, Petroleum GeoScience, vol. 3. EAGE, Amsterdam.

Deutsch, C.W., Journel, A.G., 1998. GSLIB Geostatistical Software Library and User's Guide. Oxford University Press, Oxford, England.

Haldorsen, H.H., Damsleth, E., 1990. Stochastic modeling. JPT 42 (4).

Liner, C.L., 2004. Elements of 3-D Seismology. Pennwell Books, Tulsa, Oklahoma.

Lines, L.R., Newrick, R.T., 2004. Fundamentals of Geophysical Interpretation. Society of Exploration Geophysicists (SEG), Tulsa, Oklahoma.

Robertson, J.D., 1989. Reservoir management using 3-D seismic data. The Leading Edge (TLE) 8, 25–31.

Yilmaz, O., 2001. Seismic Data Processing, Investigations in Geophysics. SEG Publications, Tulsa, Oklahoma.

Fundamentals of Petroleum Geology

2.1 INTRODUCTION

Petroleum engineers are responsible for planning and executing the development and production of petroleum reserves. In most cases, however, they are usually not heavily involved with the discovery, delineation, and evaluation of new oil and gas fields. Those tasks are normally carried out by the geologists, geophysicists, and petrophysicists of an "Operating Company" in its Exploration and/or Development Department.

In this chapter, we provide a brief overview of petroleum geology. We begin with the formation of organic matter and the origin of petroleum. We then discuss occurrence of petroleum systems, comprised of Source Rock, Burial Depth and Temperature, Reservoir Rock, Migration Pathways, Reservoir Seals, and Traps. Figure 2.1 depicts a petroleum system with its

Developments in Petroleum Science, Vol. 60. http://dx.doi.org/10.1016/B978-0-444-50662-7.00002-0

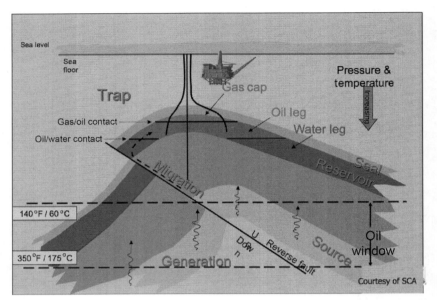

FIGURE 2.1 A petroleum system. (For color version of this figure, the reader is referred to the online version of this chapter.)

different components. We elaborate on different types of petroleum traps such as structural, salt related, and stratigraphic traps. We then discuss various types of reservoir rocks such as clastic (sandstone and shale) and carbonate rocks.

We conclude with geology, geophysics, and petrophysics, in connection with reservoir geometry, volume, and assessment of reserves. In this section, we discuss how geology combined with geophysical techniques defines the geometry of a petroleum reservoir and how petrophysics is utilized to quantify the reservoir quality and petroleum reserves.

2.2 FORMATION OF ORGANIC MATTER

With the notable exceptions of certain astronomers, most scientists, in the petroleum industry, contend that petroleum resources are primarily organic, in origin.

Certain types of organic matter formed at the Earth's surface eventually produce hydrocarbons. The process starts with photosynthesis in which plants convert water and carbon dioxide to complex sugars (glucose) using the energy of the sun. Glucose is the starting material for the synthesis of more complex organic compounds either in plants or the animals that eat them. Generally, most of the organic matter produced by photosynthesis is eventually returned to the atmosphere as carbon dioxide. Only about one CO_2, molecule in every million taken up by photosynthesis is converted to hydrocarbons. This recycling of CO_2 is achieved by plant and animal respiration and through oxidation and bacterial decay when organisms die. However, the recycling of carbon as CO_2 is not totally efficient in that a very small amount (about 0.0001%) escapes and is buried.

Sediments, laden with dead (plant and animal) lake, or sea organisms are heavier than water, and naturally deposit in the lower areas or basins under the sea. These basins are originated by tectonic action and sea level changes. When the sea level rises (relative to the base of a depositional basin), the sediments are buried deeper. As ocean basins gradually fill with layers of sediments, the weight of the newer layers increases the pressure on the layers below. This weight, or pressure at depth, along with heat, converts the organic material to oil and gas.

The primary source of the organic matter that is ultimately transformed into oil and gas are the remains of phytoplankton; microscopic floating plants such as diatoms. The best environment for the accumulation of this organic matter is in quiet waters such as a swamp, lake, or deep ocean basin. Here, the organic matter can lie buried without being disturbed. However, to ensure its preservation and to prevent rapid decay, the water conditions need to be stagnant and reducing (oxygen deficient or anaerobic) thus eliminating the possibility of aerobic bacterial decay or scavenging by fish, etc. Along with the organics, muddy sediment also accumulates. Source rock starts life as an organic-rich mud, subsequently to be converted to a claystone, shale, or marl.

2.3   ORIGIN OF PETROLEUM

Petroleum hydrocarbons are complex substances formed from hydrogen and carbon molecules and sometimes containing other impurities such as oxygen, sulfur, and nitrogen. They come in many combinations and types, from the petroleum products used in cars and other internal combustion engines to natural gas used for heating and cooking. There are "light oils" and "heavy oils," wet gas and dry gas. However, what they all have in common is an origin from organic matter; that is plants and small animals that were once alive that have created the "source rock."

"Source rocks," the rocks that produce hydrocarbons, are rich in particular types of organic matter. Chemical changes after burial convert plant and animal tissue to the complex molecules that eventually produce oil or natural gas by the effects of heat and pressure on sediments trapped beneath the Earth's surface over millions of years. The ancient societies in Egypt, China, and India made limited use of petroleum mainly as fuel for lamps, medicine, and as caulking for boats and canoes. The modern petroleum age began a century and a half ago. Advances in technology have steadily improved our ability to find and extract oil and gas and to convert them to efficient fuels, lubricants, and other useful consumer products.

2.4 OCCURRENCE OF PETROLEUM SYSTEMS

Petroleum systems occur in reservoirs within sedimentary basins—those areas of the world where subsidence of the Earth's crust has allowed the accumulation of thick sequences of sedimentary rocks. Petroleum is composed of compressed hydrocarbons and was formed millions of years ago in a process that began when aquatic plant and animal remains were covered by layers of sediments (particles of rock and mineral). As bacteria and chemicals broke down the organic plants and animal material, increasing layers of sediment settled on top. Heat and pressure transformed the layers of sediment into sandstone, limestone and other types of sedimentary rock, and transformed the organic matter into petroleum. Tiny pores in the rock allow the petroleum to seep in. These *reservoir rocks* hold the oil like a sponge, confined by other, low permeability layers that form *traps*. For a rigorous definition and more on petroleum systems see, for example, Magoon and Dow (1994).

2.5 SEDIMENTATION AND DEFORMATION PROCESS

Before we get into petroleum reservoirs, it is important to discuss the sedimentation process with which much of the oil and gas reservoirs are associated. We also want to briefly discuss different geologic time periods, how different geologic structures are formed and how such structures have evolved over millions of years. Figure 2.2 shows a picture of a geologic formation that is visible to us (outcrop). A typical outcrop, such as the one depicted in Fig. 2.2, contains a vast amount of information about many different tectonic movements, sedimentation processes, uplifting, subsidence, deformation, and other evolutionary natural events that geologists can uncover such historical events through various types of modeling and testing of the hypothesis. Subsequent chapters show how geophysical data can help geologists with their analysis in building more reliable models.

The evolution of formation of geologic structures is accomplished through careful analysis of natural processes, modeling, and various

FIGURE 2.2 A typical outcrop demonstrating different rock formations and starta. *Black Dragon Canyon, Utah Photo by Fred Aminzadeh.* (For color version of this figure, the reader is referred to the online version of this chapter.)

hypotheses. For example, Fig. 2.3 shows different time frames in which a "disconformity" is formed. At the top of the figure, going back several million years, the sediments characterized by ABCD were deposited under the sea bed. Then (the second model from the top), the uplift of the beds above the sea level, caused by tectonic forces, expose them to erosion. Note the erosion has stripped away sediment package D and part of C, creating an irregular collection of hills and valleys (model 3 from above). Finally, at the bottom model, we note creation of a new package of sediments marked as E created from subsidence below the sea that is deposited on top of C. The irregularity of C package is preserved as an unconformity. Later on, we will see (e.g., when we discuss seismic attributes) how seismic data and petrophysical information can help resolve different sediment packages (chronostratigraphic units) and the corresponding unconformities.

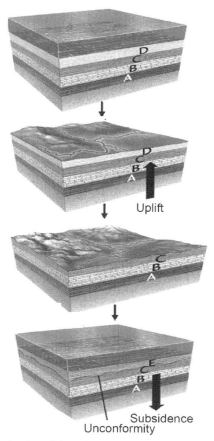

FIGURE 2.3   A Schematic view of the deformation, uplift and subsidence process. *From Levin (2013). Reprinted with permission of John Wiley & Sons, Inc.* (For color version of this figure, the reader is referred to the online version of this chapter.)

Similarly, different stages in the process of formation of the angular conformity are demonstrated in Fig. 2.4.

Figure 2.5 shows another outcrop based on which geologists would interpret and come up with a plausible assessment of how sedimentary breaks, or "disconformities," have evolved. Here, we can see the flat layers of rock that at first glance look like continuous layering of sediment. The two

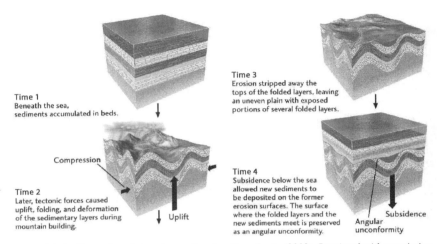

Time 1
Beneath the sea, sediments accumulated in beds.

Time 2
Later, tectonic forces caused uplift, folding, and deformation of the sedimentary layers during mountain building.

Compression

Uplift

Time 3
Erosion stripped away the tops of the folded layers, leaving an uneven plain with exposed portions of several folded layers.

Time 4
Subsidence below the sea allowed new sediments to be deposited on the former erosion surfaces. The surface where the folded layers and the new sediments meet is preserved as an angular unconformity.

Angular unconformity

Subsidence

FIGURE 2.4　Formation of Angular conformity. *From Levin (2013). Reprinted with permission of John Wiley & Sons, Inc.* (For color version of this figure, the reader is referred to the online version of this chapter.)

Disconformity

20 million years

150 million years

FIGURE 2.5　A picture of Grand Canyon depicting the unconformities and the associated geologic time. (For color version of this figure, the reader is referred to the online version of this chapter.)

formations highlighted are associated with the 20 million year gap between the "Redwall" and the "Supai" and the 150 million year gap between the "Muav" and the "Redwall." This can further be confirmed by looking at the associated fossils that allow us to determine the ages of the rocks and determine the large gaps in the geologic times between the corresponding layers.

2.6   GEOLOGIC TIMES

It is important to recognize that geologic times are associated with different sedimentation processes over millions of years, corresponding to different "strata." While due to the development of new dating methods and refinement of previous ones, geologic time scales have gone through constant revisions, the main geologic times are somewhat well established. Figure 2.6 shows a

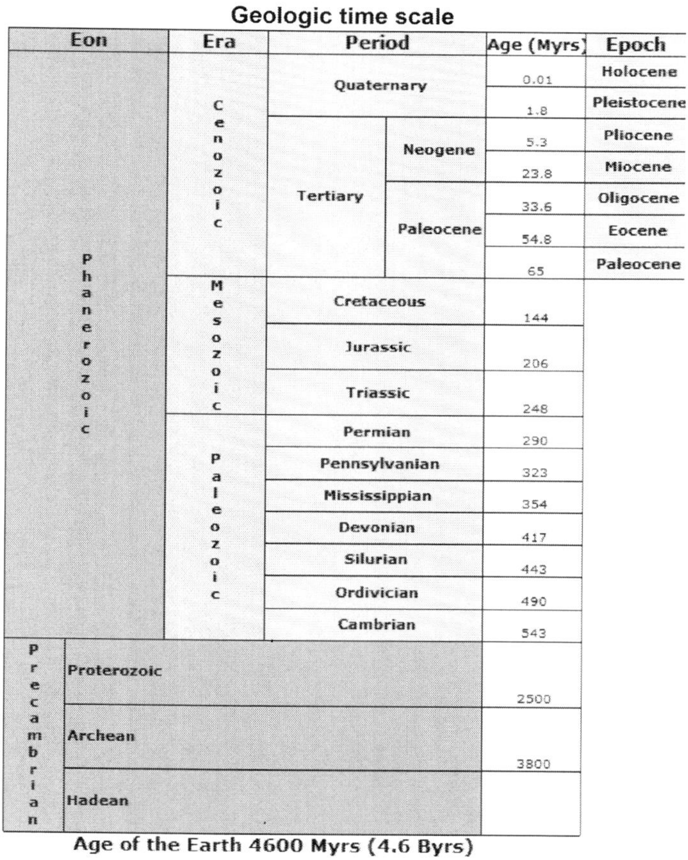

Geologic time scale

Eon	Era	Period		Age (Myrs)	Epoch
Phanerozoic	Cenozoic	Quaternary		0.01	Holocene
				1.8	Pleistocene
		Tertiary	Neogene	5.3	Pliocene
				23.8	Miocene
			Paleocene	33.6	Oligocene
				54.8	Eocene
				65	Paleocene
	Mesozoic	Cretaceous		144	
		Jurassic		206	
		Triassic		248	
	Paleozoic	Permian		290	
		Pennsylvanian		323	
		Mississippian		354	
		Devonian		417	
		Silurian		443	
		Ordivician		490	
		Cambrian		543	
Precambrian	Proterozoic			2500	
	Archean			3800	
	Hadean				

Age of the Earth 4600 Myrs (4.6 Byrs)

FIGURE 2.6   A schematic view of geologic times. (For color version of this figure, the reader is referred to the online version of this chapter.)

typical geologic timetable: the approximately 4.6 billion years of the Earth's life span are divided into major intervals or geologic periods. For example, much older formations (e.g., pre-Cambrian) are at the bottom of the geologic time with an age range starting from 550 million years. The "younger" sediments (e.g., those from Cretaceous period, belonging to the Mesozoic era) with an age range of 65–144 million years (aka Myrs) are near the top of the geologic age range.

2.7   PETROLEUM RESERVOIRS

Oil and gas accumulations are result from the coincident occurrence of the following six elements:

Source Rock
Burial depth and temperature
Reservoir Rock
Migration pathways
Seal Rock
Trap

Three-dimensional (3D) seismic surveys enable the geologist and geophysicist to investigate many of these key elements—identifying likely migration paths, inferring the relative timing of trap formation and charge and measuring the geometry and size of closed structures. In some cases, reservoir quality and even the presence of fluid hydrocarbons may be estimated. Rock physics is a key component of analyzing the reservoir. Much of this is related to the source rock, seals and the capacity of the reservoir to contain hydrocarbons. Porosity is a key ingredient and will determine the supply of petroleum that is contained in the rocks. Seismic velocity can be related to porosity. The empirical Wyllie Time Average equation (Wyllie et al., 1956, 1958) relates velocity to porosity by using the time average of acoustic (seismic) travel through the rock matrix and the fluid-filled pores.

$$\varphi_{wyl} = \frac{1}{\mathbf{B}_{cp}} \left(\frac{\Delta t - \Delta t_{ma}}{\Delta t_f - \Delta t_{ma}} \right), \tag{2.1}$$

where φ_{wyl} is the Wylie Time Average sonic porosity. Δt is the observed interval transit time (inverse velocity). Δt_{ma} is the matrix (solid) interval transit time. Δt_f is the fluid interval transit time. $\mathbf{B}_{cp}$ is an arbitrary constant used to keep the Δt–φ relationship linear.

Wyllie's equation is reasonably valid for consolidated sandstones but is generally an oversimplification for unconsolidated sandstones and carbonates. As a general rule, velocity decreases will accompany porosity increases, as related by another empirical relationship by Raymer et al. (1980):

$$\varphi_{RHG} = RHG\left(1 - \frac{\Delta t_{ma}}{\Delta t}\right), \tag{2.2}$$

where φ_{RHG} is the Raymer–Hunt–Gardner sonic porosity. Δt and Δt_{ma} are as above. RHG is an arbitrary constant ($0.4 < RHG < 0.8$).

For the formation of hydrocarbons within the basin: there must have been a *source rock*, rich in organic carbon (a rock with abundant hydrocarbon–prone organic matter 0.5–2% by weight).

For conversion of organic matter to hydrocarbons, there must have been sufficient heat over long periods of time to convert the organic carbon into hydrocarbons. The temperature for oil generation and maturation is 50–150 °C. Such high temperatures are usually achieved at depths of between 2 and 4 km. Thus the sedimentary basin will need to be deep enough to ensure that the source rock reaches the required depth.

2.8   HYDROCARBON RESERVOIRS

Hydrocarbon reservoirs are rocks that have:

- Sufficient *porosity* (void space) to store commercial volumes of hydrocarbons.
- Sufficient *permeability* (fluid flow capability) to be able to deliver the hydrocarbons to extraction wells.
- Sufficient *hydrocarbon saturation* (volumes of hydrocarbons relative to other fluids) to be an economic resource.

Since oil is lighter than water and gas is lighter than both, when a hydrocarbon reservoir is found, it is stratified with gas on top, oil in between, and water on the bottom, if all three phases are present.

Sedimentary rocks fall into one of four basic groups. These are sandstones, shales, carbonate rocks, and evaporites. These rocks are generated by two principal processes:

1. Erosion, transport, and deposition of sediments, as well as
2. Chemical solution and precipitation.

The erosion process is one in which solid particles resulting from land weathering are transported and usually deposited in water environments as sediments. The solid materials result from complete weathering of igneous rocks. Sediments accumulate as fragmented material and result in sedimentary deposits having a clastic texture. As the sedimentary material is transported, abrasion processes round the grains.

2.8.1   Clastic Sediments

Clastic sediments are predominantly clay minerals and quartz particles, with minor amounts of Feldspars, micas, and heavy minerals. Porosity results from the space between the grain particles that is not filled with cement or clay. Porosity is usually in the range from 10% to 30% depending on the grain sizes, compaction, and the amount of cement present between the pores. Permeability, which is the property that permits fluid to flow through the pores, is controlled by the amount of cement, the degree of compaction, and the magnitude and variation of grain sizes.

2.8.2 Chemical Sediments

The second source of sedimentary deposits is a result of chemical precipitation of solids from solution in water. Dissolved solids in surface waters also are weathering products. Soluble salts are leached from rocks during weathering and transported by flowing waters to quiet waters where they are precipitated, by either organisms or evaporation. The Colorado River, source of much irrigation and drinking water in Southern California and Arizona, is notable in its unusually high dissolved solids content.

Limestones are formed by chemical precipitation of calcite (calcium carbonate) or by aggregation of preexisting calcite particles. The most common sources of these particles are animal skeletons and plant secretions. The soluble elements precipitating from water are primarily calcium, magnesium, sodium, potassium, and silicon. The bulk of this chemical precipitation is organism produced and referred to as biogenic chemical sediments. Nonbiogenic chemical sediments are much less common and result from evaporation. Chemical sedimentation results in a granular texture. The size of the particles greatly influences the porosity and permeability of limestones. Extremely fine particles result in a very dense, low permeability rock termed micrite.

Water flowing through the pores of a limestone can greatly change the texture of the rock by leaching the grains to produce vugs. If interconnected, these vugs result in locally high porosity and very high permeability. Both the porosity and permeability of a limestone can be reduced if more calcite is deposited. Dolomite may be formed if the water causes partial replacement of calcium by magnesium to form magnesium-calcium carbonate. Such post depositional chemical process is known as diagenesis. Dolomitization results in increased porosity because the dolomite crystal unit cell is more compact than that of calcite, which it replaces. The porosity and permeability of a bed can be greatly influenced by the degree of dolomitization. Porosities of carbonate rocks range from 5% to 35% to even large caverns such at the Yates field, in West Texas.

2.8.3 Source Rocks

Not all sedimentary rocks contain oil or gas. Oil and natural gas originate in *petroleum source rocks*. Source rocks are sedimentary rocks that were deposited in very quiet water, usually in still swamps on land, shallow quiet marine bays, or in deep submarine settings. Source rocks are comprised of very small mineral fragments. In between, the mineral fragments are the remains of organic material, usually algae, small wood fragments, or pieces of the soft parts of plants and animals. When these fine-grained sediments are buried by deposition of later, overlying sediments, the increasing heat and pressure resulting from burial turns the soft sediments into hard rock strata. If further burial ensues, then temperatures continue to increase. When temperatures of the organic-rich sedimentary rocks exceed 120 °C (250 °F), the organic remains within the rocks begin to be "cooked" and oil and natural gas are formed from the organic remains and expelled from the source rock. It takes millions of years for these source rocks to be buried deeply enough to attain these *maturation temperatures* and additional millions of years to cook (or generate) sufficient volumes of oil and natural gas to form commercial accumulations as the oil and gas.

Petroleum is composed of hydrocarbons and was formed millions of years ago in a process that began when aquatic plant and animal remains were covered by layers of sediments (particles of rock and mineral). As bacteria and chemicals broke down the organic plants and animal material, increasing layers of sediment settled on top. Heat and pressure transformed the layers of sediment into sandstone, limestone and other types of sedimentary rock, and transformed the organic matter into petroleum. Tiny pores in the rock allowed the petroleum to seep in. These "reservoir rocks" hold the oil like a sponge, confined by other, very low permeability layers that form a "trap."

2.9   PETROLEUM TRAPS

Most hydrocarbon molecules are lighter than water and unless impeded, they rise toward the surface. For commercial accumulations of hydrocarbons in reservoir rocks, there must have been migration pathways or avenues, in rocks, through which hydrocarbons migrate from the source rock and, reach a trap. A hydrocarbon trap is some geometrical configuration of very low permeability rocks (seals) in configurations, which halts further migration.

A seal rock keeps the oil entrapped in reservoir from migrating away. Understanding the geological and geomechanical nature of seals has become one of the vital issues for successful exploration and development efforts. Seal rocks are very low permeability formations through which oil and gas cannot move effectively—such as mudstone, silt, clay stone, and anhydrite.

There are many types of hydrocarbon trap mechanisms. Figure 2.7 shows different types of trap with their associate geologic features.

We will highlight the following four basic forms of traps in petroleum geology further:

- Anticline Trap
- Fault Trap
- Salt Dome Trap
- Stratigraphic Trap
- Traps with the Fracture Network.

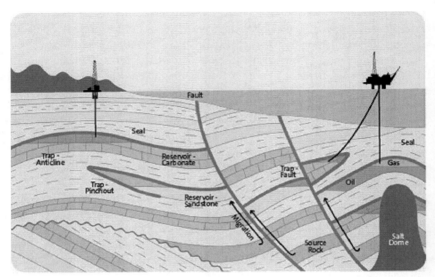

FIGURE 2.7 Display of different types of hydrocarbon traps seals, source rocks and migration paths of hydrocarbons generated in the source rocks. *Courtesy of BG-Group. http://www.bg-group.com/OURBUSINESS/OURBUSINESS/Pages/GeologyandGeophysics1.aspx.* (For color version of this figure, the reader is referred to the online version of this chapter.)

The common link between the first three is simple: some part of the Earth structures has moved in the past, creating a barrier (seal) to hydrocarbon flow. Stratigraphic traps are result of sedimentation process.

2.9.1 Anticline Trap

An anticline is an example of rocks, which were previously flat, but have been bent into an arch. Hydrocarbons that find their way into a reservoir rock that has been bent into an arch will flow to the crest of the arch, and get stuck (provided, of course, that there is a seal rock above the arch to keep the entrapped hydrocarbons in place).

Figure 2.8 is a cross section of the Earth showing typical Anticline Traps. Figure 2.9 shows a cross section of the seismic image of an anticline trap. Reservoir rock that is not completely filled with oil also contains large amounts of salt water. In most cases, such reservoirs also include a "gas cap" with the associated gas forming under a seal on the top of the reservoir.

2.9.2 Fault Trap

Fault traps are formed by movement of rock along a fault line. In some cases, the reservoir rock has moved opposite a layer of impermeable rock. The

Typical Anticline Formation

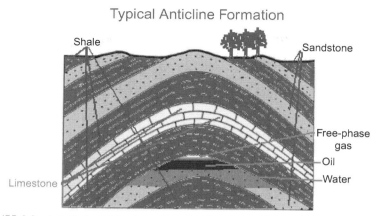

FIGURE 2.8 A typical anticline trap. From the Department of Natural Resources, Lousiana State Government: http://dnr.louisiana.gov/assets/TAD/education/BGBB/4/oil_anticline.gif. (For color version of this figure, the reader is referred to the online version of this chapter.)

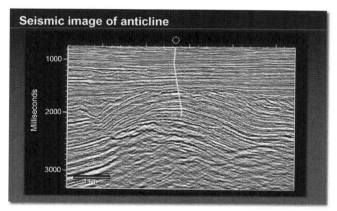

FIGURE 2.9 Seismic imaging of the subsurface, showing an anticline trap. (For color version of this figure, the reader is referred to the online version of this chapter.)

impermeable rock thus prevents the oil from escaping. In other cases, the fault itself can be a very effective trap. Movement along the fault surfaces generates a very fine-grained "rock flower," or "gauge" within the fault zone which is smeared as the layers of rock slip past one another. This very fine-grained material has such low permeability that it can act as a trap to prevent further hydrocarbon migration.

Figure 2.10 shows a cross section of rock showing a fault trap—in this case, an example of *gouge*. This is because the reservoir rock on both sides

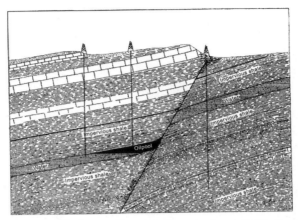

FIGURE 2.10 A typical Fault trap. From http://www.kgs.ku.edu/Publications/Bulletins/6_1/ 02_origin.html. (For color version of this figure, the reader is referred to the online version of this chapter.)

of the fault would be connected, if not for the fault separating the two. In this example, it is the fault itself that is trapping the oil.

2.9.3 Salt Dome Traps

Salt domes and diapers buried kilometers below the surface of the Earth move upward until they break through to the Earth's surface, where they are then dissolved by ground- and rainwater. In the subsurface under heat and pressure, salt deposits will flow, plastically, much like a glacier that slowly but continually moves up. To get all the way to the Earth's surface, salt has to push aside and break through many layers of rock in its path. This is what ultimately will create the oil trap.

As is shown in Fig. 2.6, salt has moved up through the Earth, punching through and bending rock along the way. Oil can come to rest right up against the salt, which makes salt an effective *seal rock*. In the Niger Delta and other very recent rapid depositional areas, shales will also move plastically and form Shale Diapers, much like salt diapers. Also shown in Fig. 2.11 is a schematic view of a seismic survey at the top (to be discussed in detail in Chapter 3) and a well drilled through the salt body.

In recent years, through the recent advances made in seismic technology, we are also able to see many prolific subsalt reservoirs (such as those seen in the Gulf of Mexico, offshore Brazil, and offshore West Africa). This was not possible in the past due to the very absorptive nature of salt, which rapidly absorbs seismic wave energy. Figure 2.12 shows an example of a Gulf of salt structure.

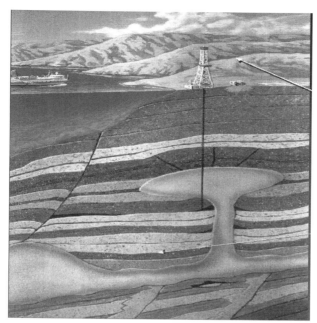

FIGURE 2.11 An example of a salt body trap. (For color version of this figure, the reader is referred to the online version of this chapter.)

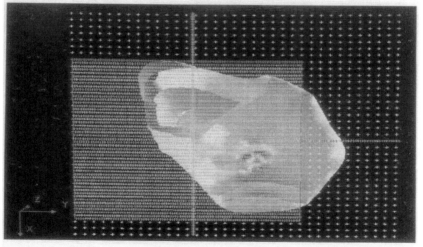

EXHIBIT 50
"3-D Salt Seismic Model"

This three-dimensional seismic model is a theoretical imitation of a complex salt dome with associated faults, sand layers and a shale sheath.

This model dimension is 8 by 8 miles. It was created to be as realistic as possible to help geophysicists better understand the data that they must properly interpret when analyzing actual salt domes in the Gulf of Mexico and elsewhere around the world. Oil and gas are commonly entrapped within sand layers that encircle or overlie salt domes.

FIGURE 2.12 A Gulf of Mexico subsalt model. *Courtesy of Union Oil/Bakersfield Muesuem of Art.* (For color version of this figure, the reader is referred to the online version of this chapter.)

2.9.4 Stratigraphic Traps

The second major class of oil trap in petroleum geology is the stratigraphic trap. It is related to sediment deposition or erosion and is bounded on one or more sides by zones of low permeability. Because tectonics ultimately controls deposition and erosion; however, few stratigraphic traps are completely without structural influence. There are many types of stratigraphic traps. Some are associated with the many transgressions and regressions of the sea that have occurred over geologic time and the resulting deposits of differing porosities. Others are caused by processes that increase *secondary porosity*, such as the dolomitization of limestones or the weathering of strata once located at the Earth's surface. Figure 2.13 shows an example of stratigraphic trap.

Stratigraphic traps are analyzed using the concepts of sequence stratigraphy which is the study of the origin, relationship, and extent of rock layers (strata). With the introduction of seismic technology, yet a newer discipline in geology was established in the sixties, called seismic sequence stratigraphy. This was aimed at utilizing seismic data to better define and understand different types of stratigraphic traps (onlap, offlap, toplap, etc.) and better map and analyze different sequence boundaries. Many seismic attributes (e.g., instantaneous phase) were introduced to better highlight different stratigraphic features (more on this in Chapter 3). Figure 2.14 shows an example of a 3D sequence stratigraphy analysis combining the seismic and well log data. The process involves transforming the conventional seismic section (in black and white) to its equivalent geologic time (through Wyler transformation) (in color). The ultimate goal of this process is to create seismic sections that are more directly related to the geologic times and the corresponding strata as described earlier.

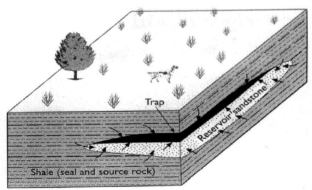

Stratigraphic trap

FIGURE 2.13 A stratigraphic trap with permeability seal.

FIGURE 2.14 Calibrating chronostratigraphy with absolute geological age using seismic and well log data. *Courtesy of dGB Earth Sciences.* (For color version of this figure, the reader is referred to the online version of this chapter.)

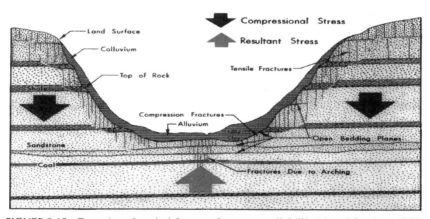

FIGURE 2.15 Formation of vertical fractures from stress relief (Wyrick and Borchers, 1981). (For color version of this figure, the reader is referred to the online version of this chapter.)

2.10 TRAPS ASSOCIATED WITH FRACTURE NETWORKS

With the recent increased interest in shale gas and liquid (oil) shale reservoirs, it is important to discuss the traps that are associated with fractures. As shown in Fig. 2.15, due to different compressional and shear stress, natural fractures are created inside the rocks. In some cases, such fracture networks (e.g., in the case of shale gas and shale oil reservoirs) become traps for hydrocarbons. Of course, in many situations processes to further stimulate these fractures to improve permeability and thus increase production, such as hydraulic fracturing, are introduced. In Chapter 9, we will further discuss this topic and introduce yet another emerging geophysical method called passive seismic or micro-earthquake (MEQ) data. We will show how MEQ data can be used both to help image and analyze the natural fractures and help design a more effective hydraulic fracturing treatment.

2.11 RESERVOIR ROCKS

Oil and gas reservoir rocks are porous and permeable (Leverson). They contain interconnected passageways of microscopic pores or holes that occupy the areas between the mineral grains of the rock. The oil and natural gas that are produced from oil and gas fields reside in porous and permeable rocks (reservoirs) in which these liquids have collected and accumulated throughout the vast expanse of geologic time.

Porosity is a measure of the spaces within the rock layer compared to the total volume of rock. Porosities are measured in percentages with the average reservoir ranging from 7% to 40%. Though both are porous, a sponge is much more porous than a brick, and though both can hold water in their pores, the sponge has a much higher capacity for holding liquids. The net rock volume multiplied by porosity gives the total pore volume: that is, the volume within the sedimentary package that fluids (hydrocarbons and water) can occupy. Figure 2.16 shows an example of a porous reservoir rock.

Permeability is a measure of how well liquids and gases can move through the rock and, thus, is a function of how well the pores within the rock are connected to each other. It is measured in units named Darcy (D). Typical reservoir permeabilities range from μD to tens of D and can vary throughout each reservoir, depending upon the type of reservoir and its origin.

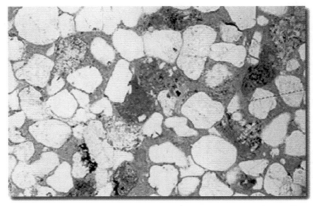

FIGURE 2.16 Photomicrograph of reservoir rocks showing the porosity and connected pores (i.e., permeability). (For color version of this figure, the reader is referred to the online version of this chapter.)

2.12   THE ROLE OF A GEOLOGIST

Geology is mostly an observational and intuitive science. The geologist works much like a detective, arriving at the scene of a crime millions of years after it was committed, with many of the clues having been destroyed and/or altered over time. Based on incomplete and sparse observations, the geologist must create a viable model of the subsurface, consistent with geologic principles, upon which an exploration and/or development program can be based. Many of the observations available to the geologist can be made by direct surface observations or airborne and/or satellite images. Figure 2.17 shows an air-photo mosaic of the breached Circle Ridge Anticline, Wyoming. Erosion has stripped off the overburden and upper layers of the structure, exposing the core and allowing geologists to infer what lies below.

Using surface-based studies of geologic relationships between the modern environments to interpret the subsurface? The present is the key to the past. Uniformitarianism and Hutton's principles are not perfect to help geologists make informed interpretations about relationships and predictions about subsurface systems. So, while we are often faced with the challenge of incomplete and discontinuous data about a specific subsurface system, we can also leverage our knowledge about geologic systems and relationships that control how lithologies are produced, structural elements evolve, and fluids and other pore-filling media evolve over time to predict and interpret what occurs in the subsurface.

Figure 2.18 shows an elevation map based on geologic interpretation of the surface geology, well control (subsurface geology), and seismic data. Such geologic maps were especially useful before the age of 3D visualization where geologists could mimic a 3D earth on a 2D map where different

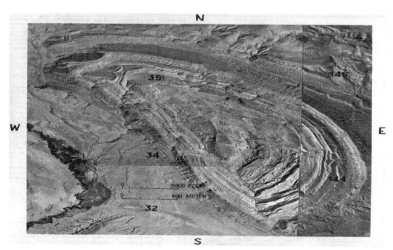

FIGURE 2.17   Air-photo mosaic showing surface of circle ridge anticline. *After Landes (1970). Courtesy of John Wiley and Sons.* (For color version of this figure, the reader is referred to the online version of this chapter.)

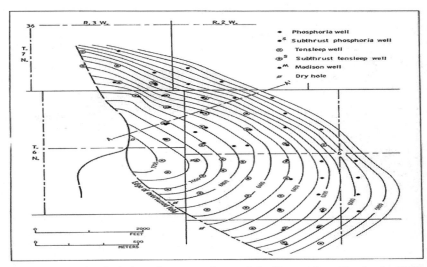

FIGURE 2.18 Phosphoria formation structural map, circle ridge anticline. *After Landes (1970). Courtesy of John Wiley and Sons.*

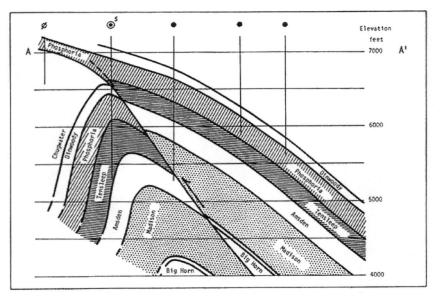

FIGURE 2.19 SW–NE structural cross section through the circle ridge anticline. *After Landes (1970). Courtesy of John Wiley and Sons.*

"contours" represent different "elevations." Naturally, the contours are truncated when a geological fault is encountered.

   Figure 2.19 shows an SW–NE structural cross section through the anticline, based upon surface geology, well control (subsurface geology), and seismic data.

2.13   THE ROLE OF GEOPHYSICS

Other geologic observations must be interpreted via indirect geophysical measurements. Geophysical measurements are Earth physical property measurements via satellite, airborne, surface, marine, and/or borehole instrument packages.

While marine, surface, and airborne gravity and magnetic surveys are often utilized for regional studies, the most commonly utilized surface and marine geophysical techniques are seismic reflection seismology. In this technique, seismic (acoustic) echoes from explosive or vibratory, and/or acoustic sources are utilized to develop subsurface models for geologic interpretation. This technique allows the geologist to image features, in three dimensions, which cannot be directly seen, much like ultra sound imaging allows physicians to noninvasively image the human body and/or materials and structural engineers to nondestructively image the interiors of complex structures.

The structural contour map of Fig. 2.18 and structural section of Fig. 2.19 are based on subsurface (well control) geology and seismic data, as well as surface geology.

2.14   EXPLORATION AND APPRAISAL WELLS

An exploration well drilled on acreage with no known petroleum reserves or production is known as a *Wildcat Well*. This name traces its origin to the early days of the oil industry, when promoters would drill wells (usually with other people's money), based upon little more than a dream, a hunch, or simply because they owned the mineral rights. The costs of drilling modern exploration wells is so great, however, that Wildcats are seldom drilled without developing detailed geological and geophysical models, first. In fact, many modern E&P organizations will spend the equivalent of the costs for several wells prior to drilling a single wildcat, because the preliminary work provides a three-dimensional model, while a well only provides information about that particular location.

If the results of a wildcat well are promising, appraisal wells are drilled to further assess the quality, distribution, and extent of the reservoir. Wireline or logging while drilling (LWD) logs are used to calculate the proportion of the sedimentary packages that contains reservoir rocks. The bulk rock volume multiplied by the net-to-gross ratio gives the net rock volume of the reservoir. Detailed evaluations of wireline logs or formation evaluation (Chapter 4) are used to estimate the amount of hydrocarbons in place.

The transition from what can be measured to what is desired requires either statistical or deterministic petrophysical models. The deterministic models, however, are really empirical models based on small numbers of laboratory measurements.

REFERENCES

Harold L. Levin, The Earth Through Time, 10th Edition, August 2013, © 2013.

Landes, K.K., 1970. Petroleum Geology of the United States. Wiley-Interscience, New York.

Magoon, L.B., Dow, W.G. (Eds.), 1994. The Petroleum System, from Source to Trap: AAPG Memoir 60.

Raymer, L.L., Hunt, E.R., Gardner, J.S., 1980. An Improved Sonic Transit Time to Porosity Transform: SPWLA 21st Annual Logging Symposium.

Wyllie, M.R.J., Gregory, A.R., Gardner, G.H.F., 1956. Elastic wave velocities in heterogeneous and porous media. Geophysics 21 (1), 41–70.

Wyllie, M.R.J., Gregory, A.R., Gardner, G.H.F., 1958. An experimental investigation of factors affecting elastic wave velocities in porous media. Geophysics 23 (3), 459–493.

Wyrick, G.G., Borchers, J.W., 1981. Hydrologic effects of stress-relief fracturing in an Appalachian Valley. U.S. Geological Survey Water-Supply Paper 2177, 51 p.

Fundamentals of Petroleum Geophysics

3.1 INTRODUCTION

Geology and geophysics are two constituent disciplines of geoscience or earth science. As we discussed in Chapter 2, geology, being an observational science, involves the study of the earth by direct analysis of rocks and formations, either from surface exposure or from boreholes, tunnels, and mines. It involves the deduction of the earth's structure, texture, composition, or history, by the analysis of such observations. Geophysics, on the other hand, is a science that deals with the physical features of the earth's surface and its internal structure. It applies the principles of physics to the study of the earth. Virtually all of what we know about the earth below the limited depths to which boreholes or mine shafts have penetrated has been derived from geophysical observations.

Historically, geophysics was principally applied in exploration—to locate the hydrocarbon-bearing reservoirs and to evaluate drillable prospects. Recently, the emphasis has shifted to the evaluation of optimum development and delineation drilling locations. This serves to maximize the start-up production rate by drilling the sweetest spots of the reservoir first. With the depletion of many oil and gas fields and the need to improve the recovery factor, geophysical technologies toward better planning for enhanced oil recovery through water flooding or steam injection are expected. In addition, horizontal drilling, especially in connection with shale reservoirs and the need for geosteering to guide the drill bit, will further necessitate the use of geophysical methods.

Geophysical data provide an accurate structure of the reservoir and a detailed characterization of the subsurface fluids and their properties. Production engineers also need the fluid saturation changes during the reservoir life cycle. Therefore, both static and dynamic characterization of the reservoir are required (Robertson, 1989). Seismic data are converted from time to depth domain using well and seismic velocity data yield the detailed structural configuration of the reservoir, including the faults that break it.

A synergism between reservoir engineering and geoscience disciplines can provide the detailed and accurate reservoir description of reservoir heterogeneity due to variations of reservoir continuity, thickness patterns, and pore-space properties. It is essential that reservoir engineering ideas and reasoning are incorporated into geoscience deductions in order for the full economic value of the data to be realized.

The *in situ* stress state of these rocks can be estimated from seismic elastic attributes. For unconventional oil and gas production from shale reservoirs, the geomechanical properties of the rocks are imperative for planning a drilling program and well completion by hydraulic fracturing. Knowledge of the stress state prior to drilling is useful for predicting areas at risk for wellbore failure. Other geophysical methods are applied in combination to corroborate the same findings as that of seismic. This characterization involves the structural and stratigraphic framework of the reservoir, its boundaries, and internal properties.

Geophysical measurements are aerially continuous and describe the pattern of distribution of rock properties over the area of investigation. When calibrated at wells, this information extends away from the well to provide 3D imaging of the distribution of physical property measured at the wells. Based on per unit data value, the cost of geophysical measurements is relatively low. With new technological innovations, the data coverage density for imaging is increasing exponentially and the operations expense in data acquisition is rapidly being reduced.

In this chapter, we will review application of geophysics methods to petroleum exploration and reservoir development. Fundamentals of geophysical techniques, their physical basis, their applications, and limitations are defined. We describe the integration of geophysical data with other data types for characterizing and predicting reservoir rock and fluid properties. More specific techniques and examples on reservoir characterization are covered in Chapter 6.

3.2 GEOPHYSICAL TECHNIQUES

There are a variety of geophysical tools and techniques that are used for characterizing oil and gas reservoirs. Geophysical tools are continuously evolving through the development of new technologies. These techniques improve subsurface imaging and increase resolution. The result is higher productivity in development wells and improved success in exploratory drilling. The higher well productivity and ultimate field recovery results from the optimal well positioning of wells by better understanding of the reservoir drainage patterns and optimizing recovery strategies based on this information.

The worldwide expenditure for seismic projects in 2011 calendar year was over $7 billion and the budget is constantly being increased every year. This represents over 70% of all exploitation and development (E&P) technology budget. The need for better data for subsurface imaging is driving this investment in seismic data acquisition. Figure 3.1 depicts a typical land seismic acquisition with the final goal of creating a 3D seismic cube. 3D seismic is by far the most used tool in seismic reflection. The largest portion of the geophysical budgets are spent on acquiring, processing, and interpreting3D seismic data. While there is a substantial amount of legacy 2D available, these data are primarily used in regional evaluation studies and is seldom used for selecting new well locations. In addition to 2D and 3D seismic reflections, there are a host of peripheral seismic techniques that are used in specific situations. These techniques include 4D time-lapse seismic, multicomponent seismic, vertical seismic profiling (VSP), and crosswell seismic.

Nonseismic geophysical techniques are also used either independently or in conjunction with seismic techniques for imaging the reservoirs. Potential field methods such as gravity and magnetics measure the ambient field and are often used for regional studies because of their low cost. However, these techniques usually do not have adequate resolution for defining new well locations. In addition, a host of electromagnetic (EM) techniques including

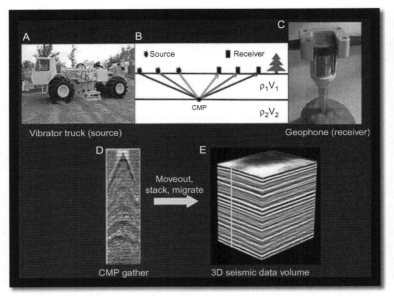

FIGURE 3.1   A typical land 3D seismic survey: (a) a seismic source: vibroseis, (b) reflected waves from a geologic boundary, (c) a geophone (seismic sensor or receiver), (d) a raw seismic record, and (e) a final 3D seismic cube or 3D seismic volume. *From Nissen (2007).* (For color version of this figure, the reader is referred to the online version of this chapter.)

DC resistivity, magnetotellurics (MT), controlled source electromagnetic methods (CSEM), and induced polarization are used in certain situations.

The interpretation of geophysical measurements along with depositional, diagenetic and other geologic data, and reservoir information and assumptions are blended together to form input parameters for the initial reservoir characterization model, (Nolen-Hoeksema, 1990). That is the subject of discussion in Chapter 6. The modeling system engine generates the integrated models that are continuously updated as new interpretation, assumptions, and data are available (Reservoir Monitoring, Chapter 7). Figure 3.2 shows the process of creation and iterative updates of common earth model.

Reservoir engineers use the estimate of oil and gas volume in the reservoir at discovery and develop a production plan for economic and efficient recovery of the hydrocarbons. Engineers forecast a production profile (rate vs. time) for the reservoir depletion and an injection profile if water or gas is to be injected. As the production proceeds, the engineers continuously assess state of the reservoir depletion and the fluid distribution by monitoring the reservoir performance and recommend remedial actions when necessary to optimize hydrocarbon recovery.

Some geophysical tools, such as seismic and EM methods, rely on using an active energy source that transmits through the subsurface rocks. The signal from the source of energy as it transmits in the subsurface is altered by the

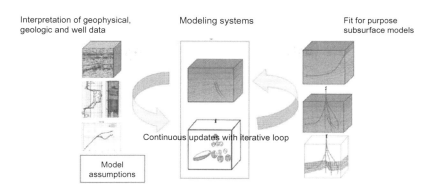

Interpretation of geophysical,   Modeling systems   Fit for purpose
geologic and well data   subsurface models

Continuous updates with iterative loop

Model
assumptions

FIGURE 3.2   A common earth model is the source for creating fit-for-purpose models. It is continuously updated by new subsurface data, interpretations, and knowledge (www.statoil.com/en/technologyinnovation). *Courtesy of StatOil ASA. Reprinted with permission.* (For color version of this figure, the reader is referred to the online version of this chapter.)

properties of rocks and this response is measured. Many other geophysical tools measure physical responses of the ambient field of rocks in a passive mode, for example, gravity, magnetics, MT, and microseismic. However, they all indirectly measure the physical properties of rocks. In most instances, these measurements are made with sensors deployed at or near the earth's surface. The measurements are made remotely from ground surface in land, ocean surface, and ocean bottom, in boreholes, airborne and from satellites. Since the measurements are made remotely, geophysical tools are generally noninvasive.

Geophysical instruments provide indirect measurement of the subsurface that can be combined with measurements such as core analysis data and well test results to better characterize the reservoir. Well data has high vertical resolution, but poor lateral definition. Correlations of well log data are often done from wells spaced from hundreds to thousands of meters apart. Geophysical data, on the other hand, are aerially sampled uniformly but have with lower vertical resolution compared to well data. Thus, the goal of geophysics is to contribute to the increment in spatial resolution that is needed for defining the building blocks of the reservoir.

3.3   INVERSION OF GEOPHYSICAL DATA

From the analysis and interpretation of geophysical measurements, models of the subsurface reservoir are derived by a process of inverse modeling. This inversion process involves integration of other different geologic and engineering data from the wells and laboratory measurements. The inverse models computed from geophysical measurements are nonunique. More than one subsurface model would fit the observations. Such nonuniqueness issues are addressed by constraining the model with additional data from other disciplines, such as geological (e.g., core data and outcrops), petrophysical measurements, as well as reservoir production data, well tests, and tracers.

The applications of the tools and the survey design are targeted to subsurface imaging for specific requirements. These are based on objectives in petroleum exploration, petroleum field development, and reservoir monitoring. Geophysical measurements, in general, are unable to directly detect the presence of hydrocarbons. Geophysicists search instead for hydrocarbon traps and assess the probability that they contain hydrocarbon reserves. After the discovery, the locations of delineation wells are decided based upon the geophysical imaging information. The data are also used to estimate the size of oil and gas accumulation in the field and for estimating reserves of the assets. The goal is to delineate the reservoir limits and assess the economic feasibility. Delineation of the geometry of reservoirs is imperative for field development. Of the four exploration risks discussed in Chapter 2, namely, source, reservoir, structure, and seal risk, geophysical methods are most effective in reducing structure and reservoir risk.

Many of the geophysical tools and techniques were originally developed for petroleum exploration. The main objective in exploration is the mapping of geological structures. These techniques are now being refined and adapted for detailed reservoir delineation and reservoir property estimation as well as for optimizing fluid production from reservoirs and for field development and drilling plans.

Geophysical methods use high-precision sensors like geophones, hydrophones, magnetometers, and gravimeters that measure specific physical properties. The small differences in physical properties that exist among various rock bodies, rather than the overall magnitudes of these properties, are what are needed. This difference in physical properties must be measured accurately. This requires high-precision sensors.

Table 3.1 lists some of the principal geophysical methods used in petroleum exploration and reservoir development and the corresponding earth property inferred from the measurement.

Geophysicists are interested in the information from the measurements that can be used to infer or interpret subsurface features. The measured attributes, by themselves, are not important. Only a limited range of subsurface conditions can give rise to a given set of surface measurements for the location. Hence, we must be able to interpret the geophysical results in terms of subsurface geology, rock, and reservoir properties.

The relative resolving power of the geophysical tools varies. This is based on the changes in the physical properties being measured in each technique. Table 3.2 shows a comparison of the relative cost, resolution, and indication of fluid from geophysical measurements.

Other techniques such as airborne gravity gradiometry and CSEM techniques are being applied in some instances for subsurface characterization for specific needs.

Geophysical measurements, inherently have limitations and ambiguities. Inversion of geophysical data is usually nonunique; this is the challenge in geophysical interpretation of subsurface geology. Techniques have been adopted to

TABLE 3.1 Geophysical Methods, Properties Measured, and Properties Interpreted

Geophysical Technique	Physical Property Measured	Earth Properties Inferred from Measurements
Seismic reflection	Travel time to acoustic boundaries, amplitude (elastic moduli and density contrast), absorption, velocity	Geological structure, depositional history, faults, rock layers, reservoir size, shape porosity, pressure, saturation distribution
Gravity	Gravitational attraction, density contrast	Geological structure, spatial variation in rock types: for example, salt domes, shale diapers, pinnacle reefs
Magnetic	Magnetic field variation, magnetic susceptibility contrast	Geometry of basement below the sediments, sedimentary cover thickness
Electromagnetic	Changes in electrical conductivity and/or permittivity	Shallow near-surface lithology changes

TABLE 3.2 Resolving Power of Some Geophysical Tools

Technique	Cost	Vertical Resolution (m)	Lateral Resolution	Fluid Discrimination
Seismic reflection	High	1–20	20–100	Poor to good
Gravity and magnetic	Low	100–1000	100–1000	Poor
CSEM	High	20–100	50–500	Excellent
Magnetotellurics (MT)	Low	100–200	100–1000	Poor

minimize them. The nonuniqueness is due to the fact that many different geological models could generate the same observed measurements. This limitation is unavoidable and arises due to fact that geophysical surveys attempt to solve a difficult inverse problem with limited resolution in the data. In spite of this limitation, however, geophysical techniques are indispensable for the investigation of subsurface geology. Main causes of uncertainty in geophysical data inversion are due to several limitations in the data. The data in all geophysical measurements limited resolution and also have random or coherent noise super-imposed on the signal. There are measurement errors in the data and significant nonlinearity of the physical process that affect the inversion. Unknown inhomo-geneity in the geology also contributes to the uncertainty in the inversion process.

A seismic reflection survey may be used to determine the depth of a buried geological interface. This would involve generating a seismic source wave at the earth's surface and measuring the travel time of the wave reflected back to the surface from a geologic interface. The conversion of this travel time into depth, however, requires knowledge of the velocity with which the wave traveled along the reflection path. If we assume a velocity, we can derive an estimate of depth but this can represent one of the many possible solutions for the depth. The velocities of propagation of seismic waves in rocks differ sig-nificantly. Therefore, it is not a straightforward matter to translate the travel time of a recorded seismic reflection horizon into an accurate depth to geologi-cal interface from which it was reflected. Even though the ambiguity inherent in the data cannot be removed, the degree of uncertainty can be reduced to an acceptable level by taking additional measurements and creating data redun-dancy. In general, the abundance and therefore the redundancy of geophysical data limit the uncertainty of the interpreted geophysical models.

The uncertainty and ambiguity associated with interpreting geophysical data could be reduced with some *a priori* knowledge of geology in the objec-tive area. Specific geologic knowledge, of the subsurface as described in Chapter 2, is invaluable in selecting the most plausible of the multiple inter-pretations from geophysical data. Integration of various measurements and their superposition, especially when well log data is available, can reduce the ambiguity in the interpretation. The general problem in geophysical surveying is that significant differences from an actual subsurface geological situation may give rise to insignificant, or immeasurably small, differences in the quantities actually measured during the survey. What we are typically interested in are not the overall magnitudes of these properties, but the small differences that exist among various rock bodies. Accuracy of measurements, therefore, relies heavily on technological development.

At borehole locations, the measured geophysical properties are calibrated with the rock and fluid properties of the subsurface layers from well data. Between the wells, the calibrated geophysical measurements are used to inter-pret lithology, porosity distribution, and fluid properties. Several methods of constraining the inversion of a dataset using data from other measurements have been suggested. Using a suite of geophysical techniques and deriving subsurface models compatible with data from other measurements reduce the nonuniqueness (Vozoff and Jupp, 1975). The most familiar technique is joint inversion to invert seismic and gravity data (Haber and Oldenburg, 1997). Musil et al. (2003) used *a priori* information from other measurements to limit the scope of inversion results.

3.4   SEISMIC REFLECTION TECHNIQUE

In seismic reflection, a controlled sound wave is generated on the ground surface or under water in marine environment and detected on the surface using instruments that respond to ground motion. As the seismic waves travel through the subsurface, they are continuously affected by the properties of the rocks. The source pulse travel through flat rock layers wave propagation gets complex as soon as any variation of the velocity and dipping beds is encountered. As seismic waves from the source travel through the earth, portions of that energy are reflected back to the surface as the energy waves traverse through different geological layers. The reflections are collected at the surface, either onshore (land surveys) or offshore (marine surveys).

Seismic reflection data (Fig. 3.3) are most commonly utilized geophysical data in petroleum exploration and reservoir characterization. The basic measurements of the seismic method are travel times that yield delay time from source to the reflectors, and the reflectivity, which can be estimated from amplitude of the recorded waves. The reflection seismic method helps to build an accurate picture of the subsurface geology. These surveys are also very cost-effective when compared to the cost of drilling a dry well. About 90%

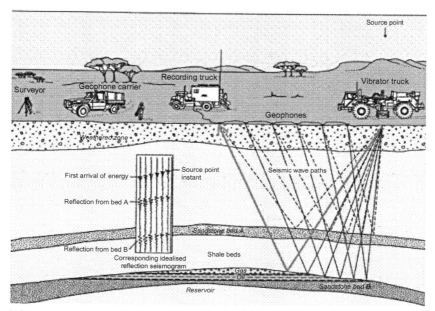

FIGURE 3.3   Seismic reflection technique showing vibroseis source, geophone receivers, and seismic recording system. The surveyor precisely marks the position of the source and receivers. The recorded field data are illustrated overlain on the figure; the raw field data is processed in the office. (For color version of this figure, the reader is referred to the online version of this chapter.)

of the effort spent on geophysical surveys for hydrocarbon exploration is in seismic methods, the remaining 10% is on the various non-seismic methods.

Seismic reflection response is a direct result of the contrast in petrophysical properties of the sediments. The acoustic wave properties completely define the wave and the rock properties necessary for determining the wave properties of a particular rock layer. The source of information on rock properties comes from laboratory measurements on rock samples and various down-hole well measurements.

Those seismic echoes or reflections are generated from interfaces between rock formations with different physical properties. The data provide valuable information about the depth and arrangement of the formations, some of which contain oil or gas deposits. The echoes from seismic explosive or vibratory sources are utilized to develop subsurface models for geologic interpretation. The echoes from which seismic reflection coefficients (R) can be estimated are related to the change in rock properties across a geologic interface. The acoustic impedance (Z), a rock property of the layer, is defined, for vertically incident waves, by the product of interval velocity (v) and density of the rock layers (ρ).

$$R = \frac{Z_{below} - Z_{above}}{Z_{below} + Z_{above}} \tag{3.1}$$

Many seismic traces (seismic section)

$$R = \frac{\rho_2 v_2 - \rho_1 v_1}{\rho_2 v_2 + \rho_1 v_1} \tag{3.2}$$

The seismic reflection as depicted in Eq. (3.1) and Fig. 3.4a is sensitive to the sequence of impedance contrasts. The collection of seismic traces comprising the traces to the right of Fig. 3.4a creates a seismic section as shown

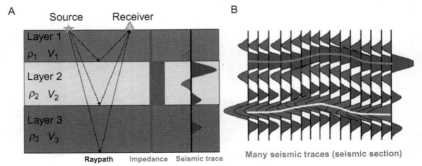

FIGURE 3.4 A schematic of a simple seismic reflection: (a) a simple three-layer earth model with the respective velocities and densities and the corresponding impedance and the resulting seismic trace and (b) the seismic section if the velocities and densities of the layer varied laterally. *From Boness (2013).* (For color version of this figure, the reader is referred to the online version of this chapter.)

in Fig. 3.4b. One way of representing the recorded seismic signal is to consider it to be composed of a source wavelet that is convolved with a series of reflection coefficients defining the rock interface, referred to as the reflectivity sequence. In an idealized experiment, this convolutional process is equivalent to the seismic source transmitting waves into the earth and the seismic receiver detecting reflected returns at the surface from each interface. The amount reflected from the lithological boundaries is defined by the reflection coefficient (R) at each interface. The reflection strength is not a measure of the physical properties of a layer, but of the contrast of properties between layers.

The amplitudes in seismic data therefore can be used for prediction of heterogeneities in reservoir rocks, net pay prediction, and fluid contacts. Lateral changes in amplitude from trace to trace along the same events or rock interface across an area could be an indicator of changes in deposition environment, porosity, rock type, or fluid saturation. The amplitude of a primary reflection is a measure of the reflection coefficient.

The seismic trace y_t as a function time is a result of reflection coefficient sequence series $R_{t-\tau}$ (fifth column from left) after conversion from depth to two-way time, convolved with the "source" wavelet w_τ, propagating through subsurface layers. Symbolically, we represent the seismic convolution as

$$y_t = \sum w_\tau R_{t-\tau} \tag{3.3}$$

The seismic trace convolution of Eq. (3.3) is the very basis of seismic reflection technology, as described by Robinson and Treitel (2009). This is illustrated in Fig. 3.5. As the seismic wave travels through the earth layers

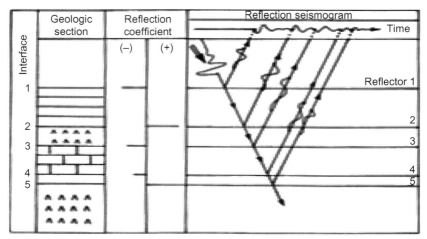

FIGURE 3.5 Convolution model of a seismic trace from reflection coefficient acoustic boundaries.

generating very small particle displacements, wave and medium act linearly so that the principle of superposition applies. Each reflected wave causes its own effect at each receiver. The total response is the linear sum or superposition of all waves from all the reflectors. From the changes in the amplitude of reflectors, the changes in the seismic impedances can be computed. The seismic impedance is used to infer changes in the properties of the rocks at the interface, such as density, porosity, and elastic modulus.

The seismic velocity of a rock layer can be expressed in terms of its elastic constants. Seismic compressional or P-wave velocity (V_p) is given by

$$V_p = \sqrt{\frac{(4/3)\mu + k}{\rho}}, \tag{3.4}$$

where μ is the shear modulus, k is the bulk modulus, and ρ is the density of the medium. Using the same notation, the shear or S-wave velocity V_s is given by

$$V_s = \sqrt{\frac{\mu}{\rho}} \tag{3.5}$$

V_p increases from air $\rightarrow$ oil $\rightarrow$ water
V_s decreases from air $\rightarrow$ oil $\rightarrow$ water

Seismic reflection techniques depend on the existence of distinct changes in the acoustic or elastic properties, seismic wave velocity, and mass–density at the subsurface rock interfaces. The energy arriving at the geophones (or hydrophone in the case of marine data) can be described as having traveled a ray path perpendicular to the wavefront. Figure 3.6 is a schematic representation of the propagation of seismic waves from the shot point source location to the reflectors in the earth and bouncing back to the surface where

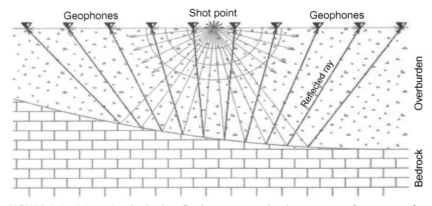

FIGURE 3.6 Schematic of seismic reflection geometry showing ray traces from source shot points to geophones. (For color version of this figure, the reader is referred to the online version of this chapter.)

the reflected energy is received at the geophone sensors. The energy spreads out as hemispherical wavefronts and is reflected from each interface or boundary of layers.

Thus, a seismic source generates acoustic or elastic vibrations on the surface that travel into the subsurface, pass through strata with different seismic responses, which alter the propagation of the waves, and return to the surface to be recorded as seismic data. As we saw earlier, acoustic impedance is the bulk density of the medium multiplied by the velocity of sound within that medium. The reflections from those boundaries are detected by the array of sensors. The signals are recorded as a function of delay time from time zero which is the initiation of the seismic source signal. The signals from deeper reflectors arrive later than from the shallower reflectors. From the seismic wave velocity in the rock, the travel time is used to estimate the depth to the reflector. For a simple vertically traveling wave, the travel time t from the surface to the reflector and back is called the two-way travel time (TWTT).

3.4.1   Compressional Waves (P-Waves)

On firing or initiation of an energy source, a compressional force causes an initial volume decrease of the medium upon which the force acts. The elastic character of rock then caused an immediate rebound or expansion, followed by a dilation force. This response of the medium constitutes a primary "compressional wave" or P-wave. Particle motion in a P-wave is in the direction of wave propagation. The P-wave velocity is a function of the rigidity and density of the medium. In dense rock, it can vary from 2500 to 7000 m/s, while in spongy sand, from 300 to 500 m/s.

3.4.2   Shear Waves (S-waves)

Shear strain occurs when a sideways force is exerted on a medium. A shear force wave may be generated that travels perpendicularly to the direction of the applied force. Particle motion of a shear wave is perpendicular to the direction of propagation. A shear wave's velocity is a function of the resistance to shear stress of the material through which the wave is traveling and is approximately half of the compressional wave velocity for dry rocks. In fluids such as water, there is no shear wave possible because shear stress and strain cannot occur in liquids.

3.4.3   Rock Physics

Rock physics is the bridge between seismic data, for example, V_s/V_p, density, elastic moduli, and reservoir properties, for example, porosity, permeability, saturation, etc. As it will be discussed in Chapter 4, petrophysics is concerned with multiple physical properties on the scale of inches, rock physics on the other hand deals primarily with physical properties effecting seismic wave propagation and on a scale comparable with seismic wavelengths or tens of feet. Rock physics uses well logs and core measurement rock P-wave velocity, density, and three-component P- and S-wave velocities to establish a relationship between the geophysical data and the petrophysical properties. We calibrate the aerially continuous geophysical data with the discrete location well measurements. This provides models for reservoir evaluation and risk assessment prior to the capital investment in drilling. We use seismic data between the well control points to resolve seal integrity issues and guide optimum placement of wells in complex reservoirs. Rock physics requires a knowledge and understanding of geophysics, petrophysics, geomechanics, and the causes of the distribution of fluids in the subsurface.

An appropriate rock physics model should be consistent with the available well, core, and seismic data. The well logs are calibrated with core data analysis. The calibrated logs are upscaled to coarser seismic scale and correlated with seismic data. This allows for a reliable prediction and perturbation of seismic responses with changes in reservoir parameters—porosity, permeability, lithology, etc., and reservoir conditions—saturation and pressure. Rock physics models can also be used in forward modeling by estimation of expected seismic properties away from the well using the observed reservoir properties at well location.

Rock physics describes a reservoir rock by physical properties such as porosity, rigidity, compressibility; properties that will affect transmission of seismic waves through the rocks. It provides connection between elastic properties measured at the surface of the earth, within the borehole environment, or in the laboratory with the intrinsic properties of rocks, such as mineralogy, porosity, pore shapes, pore fluids, pore pressures, permeability, viscosity, stresses, and overall architecture such as laminations and fractures. Description of rock and fluid properties between well control points requires understanding of the linkage of bulk and seismic properties to each other and their changes with geologic age, burial depth, and location. Figure 3.7 shows calibration of petrophysical measurements (well log data) with the rock physics (core) data.

3.5　SEISMIC WAVE PROPAGATION AND ATTENUATION

If a pebble is dropped on still water, concentric circular ripples will propagate from the location of impact or source point. As the disturbance expands about the point of impact, the amplitude of the wave (ripple) will decrease for two reasons:

1. The total energy of the disturbance created by the initial impact must be distributed around the circumference of an increasingly larger circle (geometrical spreading).

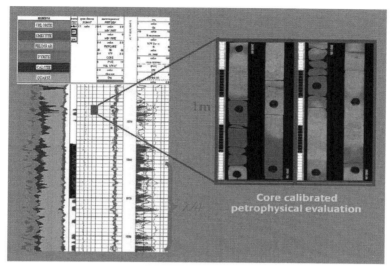

FIGURE 3.7 Calibration of petrophysical parameters of well logs with core measurements (Courtesy: http://www.senergyworld.com/technical-excellence). (For color version of this figure, the reader is referred to the online version of this chapter.)

2. The friction between the water molecules being moved by the passage of the wave will decrease (attenuate) the magnitude of the disturbance.
3. Seismic wave amplitudes are reduced as waves propagate through an elastic medium, and this reduction is generally frequency dependent.
4. Attenuation characteristics reveal much information, such as lithology, physical state, and degree of saturation of rocks.

For nonnormal incidence (as shown in Fig. 3.8), the situation becomes more complicated, as described by Mavko et al. (2003). The simple normal incidence reflection coefficient described in Eq. (3.2) has to be modified (see Eq. 3.13). Aside from a reflected compressional wave, "converted" and transmitted shear waves with different angles are generated. As shown in Figure 3.12, if the wave front ray of an incident wave intersects a discontinuity at an angle ϕ_1 to the boundary normal, then the reflected P- and S-wave rays will have angles ϕ_1 and ϕ_3 and transmitted P- and S-wave rays will have an angles r and ϕ_2, subject to the relationship referred to the Snell's law:

$$\mathrm{Sin}\,\phi_3 = (V_{s1}/V_{p1})\mathrm{Sin}\,\phi_1 \tag{3.11a}$$

$$\mathrm{Sin}\,\phi_2 = (V_{s2}/V_{p1})\mathrm{Sin}\,\phi_1 \tag{3.11b}$$

$$\mathrm{Sin}\,r = (V_{p2}/V_{p1})\mathrm{Sin}\,\phi_1 \tag{3.11c}$$

where V_{p1} and V_{s1} are the acoustic (P) and shear velocities of the layer 1. V_{p2} and V_{s2} are those for layer 2.

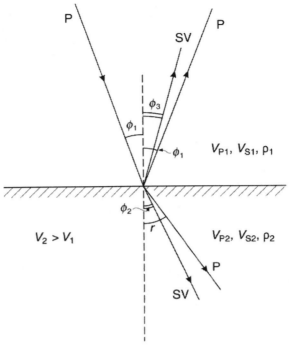

FIGURE 3.8 Schematic of a nonnormal incidence with the reflected and transmitted P- and S-waves.

The reflected P-wave in this case will not only be a function of the velocities and the densities in the two layers but also a function of shear wave velocities and the incident angle. As we will discuss later, this becomes the foundation for "amplitude versus offset" (AVO) which is a tool for detecting hydrocarbon reservoirs among other things.

3.6 SEISMIC DATA ACQUISITION

Geophysical contractors specialized in the data gathering generally perform seismic data acquisition in the field or in offshore areas. The survey parameters are carefully designed based on the knowledge of the area, the subsurface objectives within the constraints of the equipment used, and the budget available. Selection of the parameters is critical to the quality and utility of the data.

3.6.1 Land Acquisition

Seismic surveys are conducted by deploying an array of energy sources and an array of sensors or receivers in an area of interest. Figure 3.9 shows a seismic survey on land. The source of seismic waves is either an explosive which

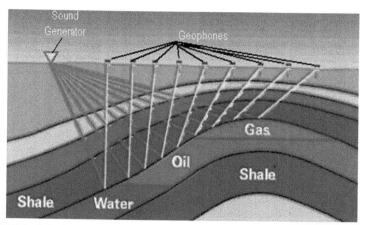

FIGURE 3.9 Seismic data acquisition on land with source and receiver locations. (For color version of this figure, the reader is referred to the online version of this chapter.)

directly generates the seismic wavelet or a mechanical source which is commonly a vibrator, which uses a steel base plate in contact with the ground and transmits seismic waves in the earth. The seismic waves with a vibrator are generated at controlled frequency ranges and a mathematical process of cross-correlation of the recorded signal with the source generated signal at the vibrator is used to create the seismic wavelet. The seismic waves that travel from the source into the earth are received on geophone sensors planted on the surface at different offsets or incremental distances away from the source point. The seismic traces are recorded as a function of time delay from the initiation of the source. For a 3D seismic survey, a network of sensors in a grid is planted and a network of source points is located. The grid of receivers and source point is moved over the survey area as the survey progresses until the entire area is covered by the survey. Each source and receiver location is surveyed for accurate surface location and elevation.

3.6.2 Marine Acquisition

Marine surveys involve (as an example) arrays of air guns towed behind the survey ship replace the vibrator or explosive source (Fig. 3.10). The implosion or collapsing of the bubble from air gun generated from high pressure compressed air constitutes the seismic source. Figure 3.11 shows marine seismic survey components. In many survey geometries, the vessel sails at a constant speed, generally approximately 4 knots (8 km along the survey profile and returns to shoot in a parallel profile after making a U-turn). The receiver sensors consist of pressure sensors (hydrophones) instead of particle motion sensors (geophones). Hydrophone receivers are towed behind the vessel in long streamer arrangements with many hydrophones mounted on the streamer.

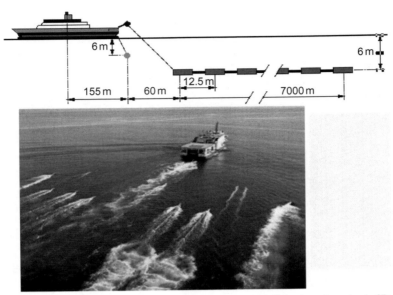

FIGURE 3.10 (Above) A typical marine survey geometry and (below) marine seismic 3D survey in progress. The survey vessel tow air guns and multiple hydrophone streamers below the water level. The recording system on the vessels records the seismic data continuously. (For color version of this figure, the reader is referred to the online version of this chapter.)

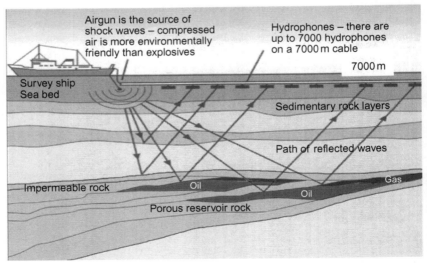

FIGURE 3.11 Seismic data acquisition in marine areas. Air guns and hydrophones are towed behind the surface ship sailing at a constant speed. (For color version of this figure, the reader is referred to the online version of this chapter.)

The air gun transmits sound waves with frequencies typically below 100 Hz through the water column and into the subsurface. Figure 3.10 shows a marine survey geometry and a real life boat in action. Figure 3.11 shows reflections from different geologic boundaries recorded.

Instead of towing receivers behind a boat, ocean bottom cables (OBCs) can be deployed by laying cables on the seafloor; OBC surveys are much more expensive compared to towed streamer survey, but do have advantages, especially for repeated time-lapse recordings.

OBCs are economically feasible in areas with obstructions and limited access where coverage is not possible for traditional towed streamers to be used. Permanently installed OBC seafloor equipment buried in trenches has also proved to be ideal and economically feasible for marine environments. Figure 3.12 shows OBC based in optical fiber cable layout and deployed on seafloor, the source array of air gun on a marine vessel sailing above the OBC grid. This is also referred to as ocean bottom network, referring to a network of geophones and hydrophones.

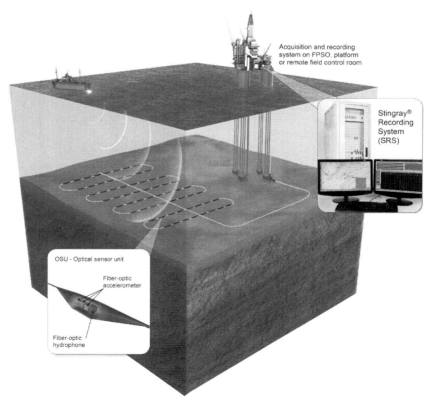

Acquisition and recording
system on FPSO, platform
or remote field control room

Stingray®
Recording
System
(SRS)

OSU - Optical sensor unit

Fiber-optic
accelerometer

Fiber-optic
hydrophone

FIGURE 3.12   A Fiber-optic Stingray® Permanent Reservoir Monitoring (PRM) system, *courtesy of TGS (www.tgs.com)*. © *TGS-NOPEC Geophysical Company ASA*. Bett, M., (2012). (For color version of this figure, the reader is referred to the online version of this chapter.)

3.7 SEISMIC DATA PROCESSING

Seismic data acquisition and processing techniques are designed to discriminate, separate, or otherwise attenuate the noise and enhance the signal. Noise reduction is a topic of research in geophysical instrument design, computer analyses, and data processing techniques. Seismic data processing converts the acquired seismic signals to a representation of the geology. Corrections for the location of the source and receivers are made and the data converted to consistent datum either at sea level or some other reference datum. The source and receiver geometry is converted to make them collocated by using a time correction based on the distance they are separated. These are known as geometric corrections. In addition to geometric corrections, data processing also consists of signal processing of the traces. This improves the signals while reducing the recorded noise in the data.

The delay between acquiring seismic data and the delivery of the final 3D seismic volume can be several weeks to months. The interpretation is only as good as the quality and accuracy of the processing of the data. Usually, the geophysical data interpreter is involved during the data processing. The main components in data processing are signal conditioning—filtering, inverse-filter or deconvolution, to extract the seismic reflectivity series for layer boundaries, correction, and compensation for acquisition geometry, common midpoint stacking and data migration for image focusing. Signal processing assumes that seismic traces are reflectivity series of the earth layers convolved with distorting filters. Deconvolution improves vertical resolution by collapsing the seismic wavelet and removes a particular type of distortion. This process provides better results when sonic logs are available, or further assumptions of the input signal are made.

Stacking uses the survey location of recorded data and uses velocity to correct for the travel delay in the distance for the source–receiver geometry, also known as normal move out correction and statics correction. The stacking or summing of amplitudes after the offset correction results in increased signal quality and significantly lowers the random noise. Either before (more expensive but more accurate) or after stacking, seismic migration is performed. Migration allows seismic events to be relocated in either space or time to the location the event occurred in the subsurface rather than the location that it was recorded at the surface, thereby creating a more accurate image of the subsurface.

The process of converting raw field records such as that shown in Figs. 3.13 and 3.14 into an image of the subsurface that is of use to engineers and geoscientists. The process can involve hundreds of different processing steps and can take several weeks of dedicated computer time to complete. Yilmaz (2001) provides comprehensive details of the different steps used to create an image of the subsurface.

Typically, the first step in seismic data processing is data "editing" to eliminate noise and selectively remove bad data. In marine seismic surveys, wave noise (swell noise) and boat traffic noise can often be problematic and can be removed with a variety of filtering approaches that take advantage of

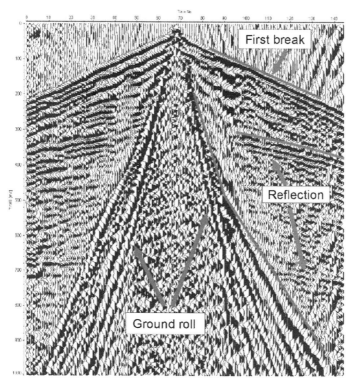

FIGURE 3.13 Seismic field record with reflection signals and noise events. Ground roll, refraction, and ambient noise in the data that need to be attenuated in processing. (For color version of this figure, the reader is referred to the online version of this chapter.)

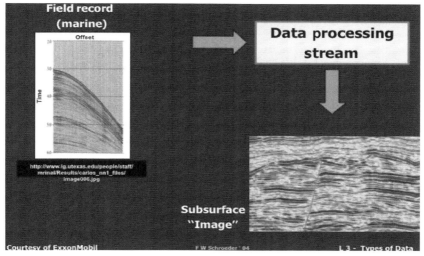

FIGURE 3.14 Seismic data processing of raw field record into image of subsurface. © *AAPG. Reprinted by permission of AAPG whose permission is required for further use.* (For color version of this figure, the reader is referred to the online version of this chapter.)

temporal and spatial differences between the noise and the desired reflection signals. On land, cultural noise from roads, power lines, buildings, and human activity, as well as ambient noise from wind and other natural causes can be attenuated with selected filtering and data editing approaches.

After data editing, seismic data are typically corrected for sensor and source effects followed by a series of processes designed to isolate primary reflection events and account for the propagation effects between the source, the receiver, and the reflecting point of interest. Propagation effects include both the loss in amplitude as the wave propagates from the source down to the reflecting horizon and the scattering and travel time effects that occur as the wave travels through different subsurface strata.

Another component of data processing is the elimination of coherent noise, also referred to as multiple suppression. Coherent noise is energy generated by the seismic source and received by the detector which is not a primary reflection. Examples of coherent noise, some of which can be observed in Fig. 3.14, are near-surface guided waves such as ground roll, non-primary arrivals such as multiples, mode conversions, and head waves. Coherent noise can be particularly difficult in land surveying. The complex weathering in the near surface can be a major instigator of coherent noise modes. Typically, coherent noise is much less troublesome for offshore surveys, and, in general, offshore surveys have higher data quality than land surveys.

3.7.1 Seismic Data Interpretation

After data processing, a 3D data volume such as that shown in Fig. 3.18 is produced. The vertical axis may be either two-way time, or if a sufficiently high quality velocity field is available, the vertical axis may be subsurface depth.

Seismic interpretation begins with mapping the structural elements of the various seismic reflectors or seismic events that are identified on the volume of processed seismic data. Some knowledge of the regional geology of the field area and specific problems in the area where the data were acquired is necessary for interpreting the data.

After the key seismic reflectors are identified, they are "picked" from trace-to-trace continuity throughout the seismic volume. Within the outline of the 3D seismic data, desired seismic horizons are picked resulting in a time structure at the top (and base) of the reservoir. This results in a structure map of the event correlated in two-way reflection time. Any disruption in the continuity could be a result of faults in the subsurface geology. Faults that disrupt the seismic horizon are marked on the horizon and picked as a fault trace. The association of these fault traces or contacts is defined. In some cases, the fault plane is mapped in the same way as a seismic horizon. The reservoir being mapped may be broken up into separate blocks by faults. Each reservoir compartment may have its own fluid contact, in which case, the compartments do not communicate with each other and need to be drained individually.

The seismic structure map in two-way travel time (also referred to as the time section) is then converted to a depth scale. This is accomplished by creating a velocity map using check shot surveys; velocity from seismic data, geologic tops from logs, and seismic time-depth conversion is a source of ambiguity in data interpretation because of the uncertainty in the velocity volume.

Interactive graphics workstations enable the interpreters to interpret 3D data volume from a 3D survey. In addition to the visualization of the subsurface, the workstation allows us to perform precise measurements of the seismic reflections that occupy the data volume. 3D visualization of computed seismic attributes for the reservoir along with geological and reservoir fluid data provide a common platform to view vast amounts of diverse data. Combining production data with time-lapse well logs, petrophysical, and seismic data and updating reservoir model and its visualization is an iterative process. Seismic attributes, such as amplitude variations, and other attributes of amplitude, velocities are displayed and mapped on the workstation. Figure 3.15 shows a typical 3D data volume that can be visualized and interpreted on a seismic workstation.

The message in the seismic amplitudes provides a measure of the reflectivity at the interfaces of rocks. Amplitudes provide information of subsurface rock property changes from the response due to the acoustic impedance. The reservoir engineers and geologists are interested in the rock and fluid properties in the reservoir interval. This is achieved by converting the amplitudes to acoustic impedance which is closer to a petrophysical property than amplitudes. This process of seismic amplitude inversion is performed by carefully calibrating the data with well logs and extracting P-wave amplitude and velocity.

Figure 3.16 shows a slice of a 3D data cube through a salt dome, and Fig. 3.17 shows a horizontal slice through the same 3D volume.

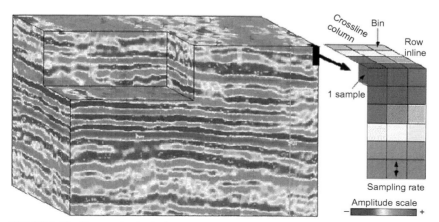

FIGURE 3.15   Processed seismic 3D volume data. The data volume represents a tight grid of stacked cross lines and in-lines. (For color version of this figure, the reader is referred to the online version of this chapter.)

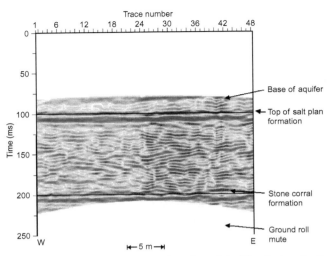

FIGURE 3.16 Vertical data slice from the northern face of the 3D seismic reflection data volume. *Courtesy of Kansas Geological Survey.* (For color version of this figure, the reader is referred to the online version of this chapter.)

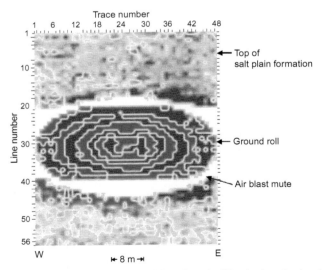

FIGURE 3.17 Horizontal data slice taken at 95 ms from the 3D seismic reflection data volume. The reflection from the salt plain formation is intersected by this data slice. *Courtesy of Kansas Geological Survey.* (For color version of this figure, the reader is referred to the online version of this chapter.)

3.8   SEISMIC RESOLUTION

Seismic resolution is the ability to distinguish separate features; that is, the minimum distance between two features so that the two can be defined separately rather than as one. The smallest interval over which a correct measurement of the distance between two closely spaced reflectors can be made is called the tuning thickness. The variation in the shape of a reflection wavelet created by closely spaced reflecting interfaces is called tuning effect.

The spatial seismic resolution measure is a wavelength. The wavelength is defined as wave velocity divided by wave frequency. In order for two nearby reflective interfaces to be distinguished well, they have to be at least 1/4 wavelength in thickness (Rayleigh criterion). This is also the thickness where interpretation criteria change. For smaller thicknesses than 1/4 wavelength, we rely on the amplitude to judge the bed thickness. For thicknesses larger than 1/4 wavelength, we can use the wave shape to judge the bed thickness. Detection is different than resolution. Although the top and bottom of a bed cannot be distinguished at a bed thickness less than 1/4 wavelength, the bed can be detected down to much smaller thickness approaching 1/20 to 1/30 of a wavelength. The detection threshold is a function of the velocity contrast of the bed with the surrounding medium.

$$\lambda = \frac{V}{f} \tag{3.12}$$

A dominant wavelength is typically observed as a function of depth. For shallow earth at upper hundreds of meters depth, $v = 1000$ m/s, $f = 100$ Hz, results in a wavelength of 10 m, while, for example, deep in the earth at 5000 m depth, $v = 5000$ m/s, $f = 20$ Hz, results in a wavelength of 250 m.

A related issue regarding seismic resolution is what is referred to as tuning thickness. The wedge model in Fig. 3.18 demonstrates how by reducing the layer thickness (from left to right) causes two distinct events associated with reflections from the top and bottom of a layer begin to interfere with each other and eventually make those events not distinguishable separately. Above the tuning gross reservoir thickness can be estimated by multiplying the time difference between reservoir top and base by the interval velocity in the reservoir. In certain situations, gross thickness can also be measured for thicknesses between the limit of resolution and tuning thickness. In cases where this applies, the gross thickness would be dependent upon amplitude strength.

There is a practical limitation in generating high frequencies that can penetrate large depths. The earth acts as a natural filter removing the higher frequencies more readily than the lower frequencies. The deeper the source of reflections, the lower the frequencies we can receive from those depths and therefore the lower resolution. A good rule of thumb and one borne out by some theoretical work by Shapiro and Hubral, (1994) is that seismic waves can only propagate coherently over distances of about 50–100 wavelengths.

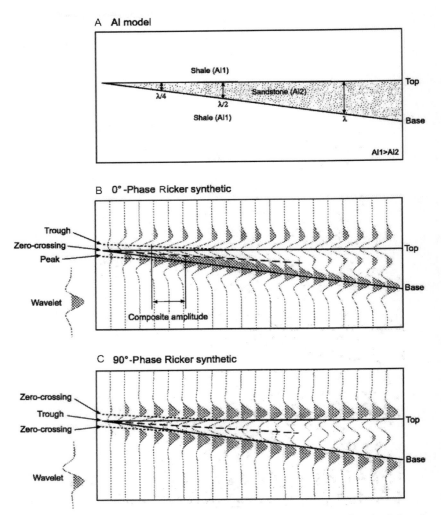

FIGURE 3.18 Wedge model showing the tuning thickness and amplitude dimming below the tuning thickness.

So, for example, if the velocity is 3000 m/s, then 50 wavelengths would be a distance of 1500 m at 100 Hz and 15,000 m at 10 Hz.

Normally, we think of resolution in the vertical sense, but there is also a limit to the horizontal width of an object that we can interpret from seismic data. This refers to how close two reflecting points can be situated horizontally, and yet be recognized as two separate points rather than one. The spatial resolution of seismic data is described in terms of Fresnel zone that defines the portion of a reflector from which reflected energy can reach a seismic

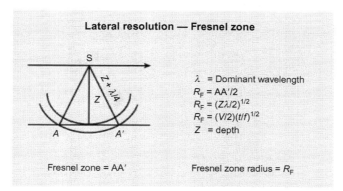

FIGURE 3.19 Effective radius of Fresnel zone is R_F where V, average velocity of the reflector; t, two-way time; f, frequency. (For color version of this figure, the reader is referred to the online version of this chapter.)

detector within half wavelength of the first reflected energy (Sheriff, 1980). The larger the Fresnel zone, the worse the spatial resolution.

A reflection is not energy from just one point beneath us; it is the energy that bounces back at us from a region. The area that produces the reflection is known as the first Fresnel zone: the reflecting zone in the subsurface is defined by the first quarter of a wavelength (Fig. 3.19). The Fresnel zone increases with depth and decreasing dominant frequency of seismic data. If the wavelength is large, then the zone over which the reflected energy returns is larger and the spatial resolution is lower. Reflected energy within this footprint contributes constructively to the recorded amplitude in the dataset. Energy from outside the footprint cancels out in the dataset. The radius of the first Fresnel zone is approximately equal to the square root of half the depth times the dominant wavelength. In theory, the process of migration that collapses the Fresnel zone (Yilmaz, 2001) can improve the lateral resolution to some fraction of a wavelength. In practice, velocity uncertainties probably limit the optimal resolution to about a wavelength of the seismic data.

3.9 SEISMIC MODELING

The "wedge model" shown in Fig. 3.18 is one aspect of what we refer to as type of "synthetic seismic data" or seismic modeling. Many aspects of seismic from data acquisition to processing, interpretation, and reservoir analysis would benefit from seismic modeling, making the results more effective and reliable. Among the reasons to create seismic models are: design a seismic survey, evaluate different processing schemes, assess the impact of changes in different reservoir properties on the seismic response, create pseudo-seismic sections from log data, test the effectiveness of different imaging

A   B   C

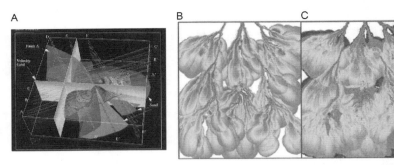

FIGURE 3.20   Seismic modeling, (a) The SEG/EAEG salt model. From Aminzadeh et al. (1997), (b) Middle Miocene sheet turbidite vshale, 35 km across. Single stratigraphic slice 20 m thick. (c) Vertical average of vshale computed over the 80 m constituting one subsheet. (b) and (c) SEG SEAM Modeling Project, From Fehler and Kelither 2011, SEEP Phase 1 Report: http://dx.doi.org/10.1190/1.9781560802945.ch2. (For color version of this figure, the reader is referred to the online version of this chapter.)

(migration) techniques, and examine the applicability of 4D seismic in a given geologic setting.

Seismic modeling, especially those used to create a large "prestack" 3D seismic response for a complex model, is very computer intensive. Two such classic synthetic seismic volumes were generated back in 1995, using the computing resources of many national laboratories with input from many oil and service companies. It is referred to as "SEG/EAEG Salt and Overthrust Models." Figure 3.20 A shows the salt structural model (with the corresponding velocity field) to generate the synthetic 3D seismic response. For details and an example dataset, see Aminzadeh et al. (1997). More recently, SEG undertook a more ambitious task generating more complicated models, including the elastic response under the project referred to as SEAM. Figure 3.20 B, after Fehler and Kelither 2011 shows a Middle Miocene sheet turbidite vshale, 35 km across. Single stratigraphic slice 20 m thick. Figure 3.20 C shows Vertical average of vshale computed over the 80 m constituting one subsheet.

3.10   SEISMIC ATTRIBUTES

Seismic attributes are computed by mathematical manipulation of the original seismic data to highlight specific geological, physical, or reservoir property features. They evaluate the shape or other characteristic of one or more seismic trace(s) and their correlation over specific time intervals. Seismic attributes computed from reflection data are based on various physical phenomena. During seismic wave propagation through earth layers, their wave characteristics, such as amplitude, frequency, phase, and velocity, change significantly. Seismic attributes provide specific quantities of geometric, kinematic, dynamic, or statistical features computed from seismic data. These changes in the seismic waves provide the signatures of the physical properties of the medium—the rocks in the subsurface through which they propagate.

An attribute-processed seismic line (in this case, calibrated impedance) that connects two well locations over the Smackover trend in W. Alabama is shown in Figure 3.21. The well on the left has porosity (and oil); the well on the right has no porosity. The black reflection cycle—a trough—is associated with the upper Smackover; on the left, it is not as black or not as negative as it is on the right, and also, on the left, there is a thin interval of white that dies out to the right. These are subtle differences that may be related to the high/low porosity. It would be hard to visually use these with only amplitude observations, but seismic attributes make it relatively easy. Another example of seismic attribute (combination of coherency and a few other attributes) is shown in Figure 3.22 where structural and stratigraphic elements are highlighted.

For more exhaustive review of seismic attributes, see Taner (2003) and Chopra and Marfurt (2007). Also, see Aminzadeh and de Groot (2006) where attributes are integrated using soft computing techniques to create indicators for different reservoir properties.

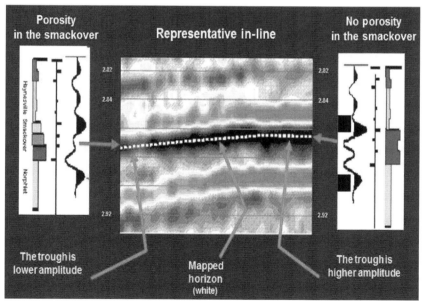

FIGURE 3.21   Seismic data from W. Alabama. Computed attributes in a section with two wells. In the left is lower amplitude that corresponds to porous Smackover measured in the well; to the right, higher amplitude trough has low porosity from the well. © *AAPG. Reprinted by permission of AAPG whose permission is required for further use.* (For color version of this figure, the reader is referred to the online version of this chapter.)

FIGURE 3.22 Multitrace blending of seismic attributes for delineation of structural and stratigraphic elements. *Courtesy of http://dGBes.com.* (For color version of this figure, the reader is referred to the online version of this chapter.)

3.11 SPECTRAL DECOMPOSITION

Spectral decomposition is a seismic attribute that has been successfully used for detecting the edges of features such as incised valleys and highlight subtle variations in the valley fill. Spectral decomposition uses the Fourier transform to calculate for each trace the amplitude spectrum of a short time window covering the zone of interest (Fig. 3.23a).

The amplitude spectrum is tuned by the geologic units within the analysis window so that units with different rock properties and/or thickness will have different amplitude responses. When the spectral decomposition is calculated for all traces in the 3D seismic volume and presented in map form (usually as a series of frequency slices), the resulting images show the lateral variability within the zone of interest.

Spectral decomposition (SD) is a powerful tool for "below-resolution" seismic interpretation and for thickness estimation. In SD, spectral properties or scale properties are extracted from a small part of the reflectivity series through mathematical transformation. As a consequence of the small transform window, the spectral response of the geological column is not "white" but contains effects such as spectral notches and tuning frequencies that relate to the local reflectivity only, hence to geological properties such as layer thickness and stacking patterns. The (combined) spectral slices highlight subtle features, often below seismic resolution, which may escape the interpreters' eye if they use the amplitude information or single attributes such as energy or instantaneous frequency alone.

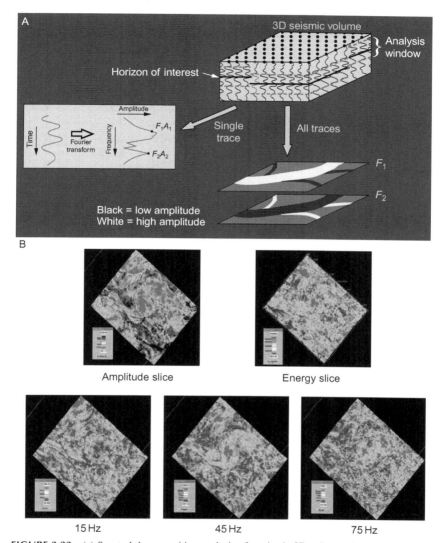

FIGURE 3.23 (a) Spectral decomposition analysis of a seismic 3D volume. (b) Horizontal slice of 3D seismic attribute volumes. *Courtesy of http://dGBes.com.* (For color version of this figure, the reader is referred to the online version of this chapter.)

An important application of SD analysis is to interpret stacked channel systems and analyze the interrelations of the different components of the channel system, from braided channel complexes, to individual channels and the channel's elements. The connectivity of different subelements can be inferred and pressure boundaries and reservoir compartmentalization can be mapped on the

SD results. If pressure data from production become available, these results can be cross-checked and refined with the pressure data from the different wells.

A meandering river system is decomposed into several spectral slices using SD. Figure 3.23 (top portion) shows the slice of a 3D seismic data volume with the amplitude and energy sections. The SD slices at 15, 45, and 75 Hz are shown at the bottom of Fig. 3.28 highlighting a meandering river system decomposed into several spectral slices. In each of the slices, we see that outside the main channel several features brighten up at different frequencies.

This gives a much better appreciation of the paleo-landscape, with smaller channels and oxbow lakes surrounding the main channel body. It is seen that within the main meander different areas brighten up at different frequencies. This may indicate variations of thickness within the channel (hence good connectivity), or maybe the channel is composed of sedimentary subbodies (possibly indicating poor connectivity), some of which may be deposited during catastrophic events (flooding), since the anomalies cross the outer boundaries as observed on the amplitude section. Some technical details could be found in Partyka et al. (1999).

3.12 ABSORPTION

High-frequency content of seismic response attenuates more extensively as it propagates through gas-bearing reservoirs. Seismic amplitude absorption is sensitive to degrees of gas saturation. This leads to an abrupt lowering of dominant frequency in the gas zone, absence of high frequencies, and dimming of reflections. Related experimental work has been published by Mavko and Nur (1979) and Batzle et al. (2006). Figure 3.24 shows the power

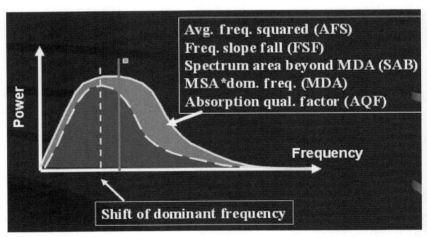

FIGURE 3.24 Fluid factor attributes relative to the power spectrum. (For color version of this figure, the reader is referred to the online version of this chapter.)

spectrum and the commonly seen shift of the dominant frequency caused by presence of gas in the reservoir.

As shown in Dasgupta et al. (2000), five fluid factor attributes have been identified, each with a specific and unique function as shown in Table 3.3. The absorption quality factor (AQF) is defined as the area of the power spectrum beyond the dominant frequency. The smaller the area, the larger the high-frequency loss and higher the probability for the wave to have traveled through a thicker gas column. An example of AQF attribute is displayed in Fig. 3.25. Note the very strong shallow absorption anomalies above 2 s TWTT, from Clifford and Aminzadeh (2011) (Table 3.4).

TABLE 3.3 Seismic Attributes for Use in Geological Interpretation

Amplitude	Lithological contrasts, bedding continuities, bed spacing, gross porosity, fluid content
Instantaneous frequency	Bed thickness, lithological contrasts Fluid content
Reflection strength	Lithological contrasts, bedding continuity Bed spacing, gross porosity
Instantaneous phase	Bedding continuity
Polarity	Fluid content, lithologic constraints
Absorption/Q	Fluid content
Coherency/similarity	Highlights faults and fractures

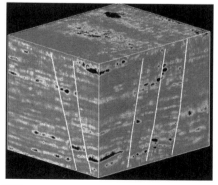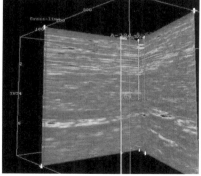

FIGURE 3.25　An example of the AQF attribute applied to the data from Grand Bay, offshore Louisiana. (For color version of this figure, the reader is referred to the online version of this chapter.)

TABLE 3.4 Fluid Factor Attributes and Their Respective Functions

Attribute	Function
Average frequency squared (AFS)	Magnifies high-frequency loss
Frequency slope fall (FSF)	Highlights flattening of spectrum
Spectrum area beyond MDA (SAB)	A measure of high-frequency loss
MSA[a] dominant frequency (MDA)	Reduces the impact of noise
Absorption quality factor (AQF)[a]	Overall measure of absorption

[a]AQF, area beyond dominant frequency weighted by frequency (see Fig. 3.24).

3.13 SIMILARITY/COHERENCY/CURVATURE

Most of the seismic attributes described above are based on the analysis of a single trace at a time. To highlight faults and fractures from a seismic data cube, other attributes that highlight lateral changes in the continuity of seismic horizons are used. One of such attribute is the coherency or similarity attribute. It is based on comparing or correlating two or more adjacent traces and to highlight lack of coherency or a measure of dissimilarity among them. The curvature attribute, mostly used to highlight smaller changes in the curvature or discontinuity in seismic reflections, is often used for highlighting fracture network in shale reservoirs. Fig. 3.26 shows a horizontal slice of these fault/fracture-related attributes.

3.14 AMPLITUDE VARIATIONS WITH OFFSET

Amplitude variations with offset or amplitude versus offset (AVO) were first introduced by Ostrander (1984). AVO is a seismic data analysis technique that uses the amplitude information of prestack seismic data as hydrocarbon

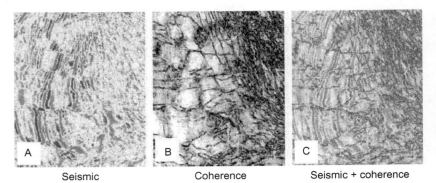

| Seismic | Coherence | Seismic + coherence |

FIGURE 3.26 Fault and fracture mapping with coherence attributes. *Courtesy of WorldOil Sep2000.* (For color version of this figure, the reader is referred to the online version of this chapter.)

indicators. At interfaces in the incident P-wave coming with nonzero angles (Fig. 3.8), a fraction of it is converted into S-waves, both transmitted and reflected. Thus, at nonnormal incidence, the PP reflection is not only a function of the acoustic impedance contrast but also a function of the S-wave velocity contrast across the interface. The AVO effect is a combination of rock physics properties of overlying lithology and the reservoir rock (VpVs and p). The impedance contrast over the top reservoir interface is the critical factor.

The AVO phenomenon is shown in (Veeken and Rauch-Davies, 2006) Fig. 3.27 where the near-offset amplitude value is different from the amplitude measured on the far-offset trace. Note the difference of the amplitude response in the water-filled or wet sand reservoir above. The changes in the petrophysical characteristics of the encasing shale sequence with depth and the diagenesis are some of the causes for the different responses. AVO provides a tool to estimate pore fill of reservoirs. The AVO analysis brings the data from the offset domain into the "Amplitude versus Angle-of-incidence" domain (AVA). The reflection coefficients at different offset and incidence angles are computed. This analysis discriminate between water and hydrocarbon saturated reservoirs.

The seismic signature from a gas or light oil (gas in solution) sand is different from the brine-filled response when the same reservoir is observed under similar conditions. In the following equation, Shuey (1985) linearized the complicated mathematics that represent the angular dependence of P-wave reflection coefficients with angle and terms A and B, the intercept and gradient. Equation (3.13) is Shuey's approximation to the Zoeppritz equations.

$$R(\theta) = R(0) + G \sin^2 \theta + F\left(\tan^2 \theta - \sin^2 \theta\right) \qquad (3.13)$$

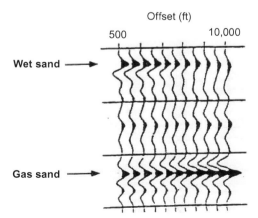

FIGURE 3.27   Amplitude variation with offset AVO effect on a flattened CDP gather caused by the presence of gas in reservoir sand.

where

$$R(0) = \frac{1}{2}\left(\frac{\Delta V_P}{V_P} + \frac{\Delta \rho}{\rho}\right)$$

and

$$G = \frac{1}{2}\frac{\Delta V_P}{V_P} - 2\frac{V_S^2}{V_P^2}\left(\frac{\Delta \rho}{\rho} + 2\frac{\Delta V_S}{V_S}\right); \quad F = \frac{1}{2}\frac{\Delta V_P}{V_P}$$

θ = average of the incident and transmitted angles; V_P = average P-wave velocity; ΔV_P = P-wave velocity contrast across interface; V_S = average S-wave velocity; ΔV_S = S-wave velocity contrast across interface; ρ = average density; and $\Delta \rho$ = density contrast across interface.

The first term of this equation or the incident angle independent part is a function of changes in density and compressional wave velocities. The second term also includes the shear wave velocity changes across the interface. Equation (3.14) is further approximated as

$$R(\theta) = A + B \sin^2(\theta) + \ldots \tag{3.14}$$

where R = reflection coefficient, θ = angle of incidence, A = AVO intercept—where the curve intersects $0°$, B = AVO gradient—a linear fit to the AVO data.

Another AVO approximation was given by Hilterman (1990):

$$R(\theta) \sim (\Delta \alpha + \Delta \rho)\cos^2(\theta) + 4\Delta \sigma \sin^2(\theta) \tag{3.15}$$

where $\Delta \sigma$ is the Poisson's ratio contrast across the interface.

Figure 3.28 illustrates AVO attributes Intercept-Gradient. Intercept (I) is cut-off from the Y-axis or amplitude Ro, Gradient (G) is slope of regression line with X-axis. A linear regression analysis is done to compute the I and G values in a 'amplitude-$\sin^2\theta$' crossplot, whereby θ = angle-of incidence. AVO brings the data from offset domain to 'Amplitude versus Angle-of-incidence' domain. I and G are cross plotted.

Earlier success of AVO applications were hampered by some unexpected drilling outcomes. This was attributed to the fact that not all geologies lend themselves to a similar AVO response. After Rutherford and Williams (1989) and Castagna and Swan (1997) and many others, four classes of AVO responses emerged. The AVO classification allows easy computation of the types (Young and LoPiccolo, 2003). By applying the classification together with information and knowledge of the geology of the area, exploration targets in geological settings and fluid contacts in developing reservoirs have been successfully identified. This characterization has resulted in a dramatic reduction of drilling risks in areas such as the Gulf of Mexico.

3.14.1 AVO Classification Standards

Class I:

- Increase in impedance
- Starts with high positive R_0 amplitude and reduces with offset

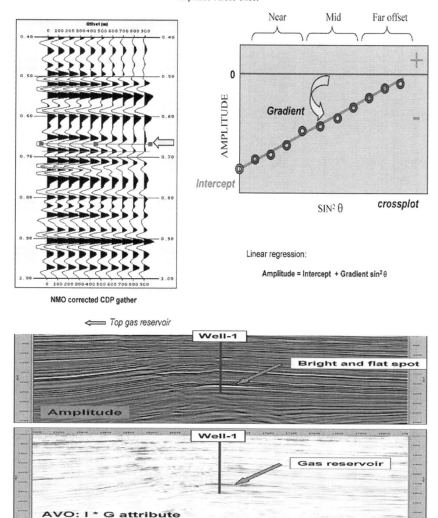

FIGURE 3.28 Crossplot of different AVO responses. For gas, oil, and water, (a) NMO-corrected CDP-gather with AVO effects. The amplitude of the near trace is different from the amplitude seen on the far offset trace. The amplitude for a particular reflection can be approximated in an Amplitude-sin$^2\theta$ plot by a straight line. The line defines the Intercept (cut-off from the Y-axis) and the Gradient (slope of the **g** line with the X-axis). (b) *(Top)* Seismic amplitude section with distinct flat spot, caused by the presence of gas in a reservoir. It also corresponds with anomalies in the Intercept-times-Gradient and Fluid Factor AVO attributes as shown in the two figures below (courtesy Pemex). *(Middle)* Intercept-times-Gradient attribute computed from the amplitude-sin$^2\theta$ crossplot. The Intercept or *Ro* is the cut-off value for the amplitude at a zero incidence angle.

(Continued)

FIGURE 3.28—Cont'd   The Gradient is the slope of the regression through the amplitude points at the different angles of incidence θ. *(Bottom)* Fluid Factor attribute is a weighted function between the Intercept and Gradient attributes. The wet rock is established in a cross plot and the distance from the individual points to this line is a measure for the Fluid Factor. *Veeken (2007)*. (For color version of this figure, the reader is referred to the online version of this chapter.)

Class IIa:

- Impedance difference is minimal
- Starts with a low amplitude and increases with offset

Class IIb:

- The same as IIa, with a polarity reversal of IIa

Class III:

- Impedance decrease
- Starts with high amplitude and increases further with offset

Class IV:

- Impedance decrease
- Starts with high amplitude and decreases with offset

The AVO classification (as shown in Figure 3.29) allows easy computation of the types. Using those along with information and knowledge of the geology of the area, attractive exploration targets in many geological settings have been successfully identified in Gulf of Mexico (Figure 3.30). This has resulted in reduction of drilling risk. The classification depends on the contrast of petrophysical properties of the reservoir rock with the overlying unit. In the case of a nonperfect top seal, the properties of the overlying unit can also change when the reservoir is gas filled (Castagna and Swan, 1997).

While AVO has proved to be a powerful tool for hydrocarbon detection (gas or light oil in particular), there are several potential pitfalls. The following highlights some of the areas where we should be careful in interpreting and using the AVO attributes: (a) impact of anisotropy in shale/sand models (Kim et al., 1993), (b) tuning effect (Bakke and Ursin, 1998), (c) acquisition and processing effects, (d) impact of complex geology, thin beds, vertical heterogeneity, faulting, and (e) scattering in heterogeneous overburden (Figure 3.30).

3.15 MULTICOMPONENT SEISMIC TECHNIQUE

The seismic system described so far is related to recording of vertical component P-wave data on P-wave receivers. Multicomponent seismic system uses three-component receivers to record elastic seismic wave and has been

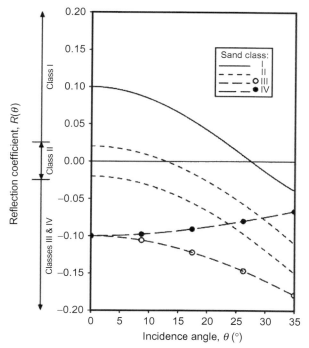

FIGURE 3.29   AVO analyses and spectral decomposition of seismic data from four wells west of Shetland, UK (Loizou, N., et al., 2008). *Courtesy of EAGE.*

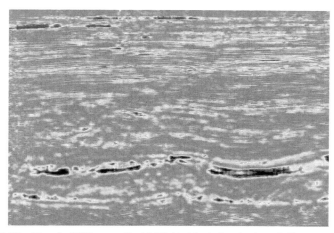

FIGURE 3.30   Class 3. AVO anomalies highlighted in dark color in a shallow reservoir in the Gulf of Mexico. (For color version of this figure, the reader is referred to the online version of this chapter.)

demonstrated to be an effective technology for risk reduction in exploration and development. The purpose of a multicomponent seismic system is to record and utilize both compressional (P) and shear (S) wave modes. Along with a vertical receiver, two horizontal receivers in orthogonal directions are added. This may lead to a greatly improved imaging compared to conventional data. The S-wave component provides seismic data that are important complementary seismic information for comparison with P-wave images and AVO results have many vital uses including lithology identification, fluid discrimination, imaging through gas, fracture and stress-field characterization, and density estimation. P-waves are influenced by all three bulk rock properties (compressibility, rigidity, and density), while S-waves are influenced by rigidity and density only. Combining these observations allows more accurate estimation of key reservoir characteristics.

Rock physics models are used in forward modeling by estimation of expected seismic properties away from the well using the observed reservoir properties at well location. Rock physics provides the link between seismic data, for example, V_s/V_p, density, elastic moduli, and reservoir properties, for example, porosity, saturation, etc. It provides a range of cost-effective technologies for evaluation and risk assessment prior to the investment in drilling. The rock physics models allow for a reliable prediction and perturbation of seismic responses with changes in reservoir parameters porosity, permeability, lithology, etc., and reservoir conditions saturation and pressure.

Additionally, multicomponent data provide geomechanical rock properties at reservoir target formations. The seismic attributes are calibrated with measurements of mechanical properties at wells and are used to extrapolate information at wellbores to the interwell regions. Lithologic and geomechanical models derived from seismic data can be correlated to predict well flow rates using multiattribute analysis. The interpretation also provides understanding of rock properties and stress characteristics at the target subsurface. The *in situ* stress state of these rocks can be estimated from seismic elastic attributes. For oil and gas production from shale reservoirs, the geomechanical properties of the rocks are necessary for drilling and well completion by hydraulic fracturing. Estimate of the stress state prior to drilling is helpful in prediction of formations that are at risk for wellbore failure.

Furthermore, multicomponent seismic records also provide additional information for reservoir characterization, improved imaging, hydrocarbon/lithology prediction, fracture and stress, as well as fluid saturation identification. The amplitude of P-waves passing through the gas is attenuated, hence obscuring deeper events. P-wave time images suffer from structural distortion caused by low velocities in the vicinity of the gas. This affects events beneath the gas giving rise to false structure. Shear wave data derived from three-component records can help resolve some of the ambiguities. This is because S-wave amplitudes are undiminished and so provide clear images under the gas. S-wave data help model these low-velocity zones to remove the push down effect from the final image. To illustrate some of these applications, we provide two case histories below.

3.15.1 Offshore Norway Multicomponent Survey

The section in Fig. 3.40 is from North Sea Snøhvit field, offshore Norway. Existing seismic data from streamer surveys suffered from poor seismic imaging, particularly in the western part of the field, caused by shallow gas. This led to uncertainties in interpretation and consequently hydrocarbon volume calculation. Multi-azimuth OBC data were recorded and processed by CGG. The multicomponent OBC results represent a significant improvement in the image of the reservoir. The results shown in Fig. 3.31 indicate improved imaging in PS seismic.

This allowed accurate interpretation of the bounding fault leading to more accurate reservoir volume calculations, and S-waves provided complementary information about the rock matrix properties, pore fluids, and pressures. From early laboratory rock physics studies to recent seismic surveys, the combined use of P- and S-wave data helps in discriminating lithology.

3.15.2 Canada Heavy Oil Multicomponent Survey

Multicomponent data have several applications in heavy oil developments, such as

- Identification of shale volume through better density estimation
- Tracking temperature changes in 4D surveys
- Identifying local variations in anisotropy

PP   PS

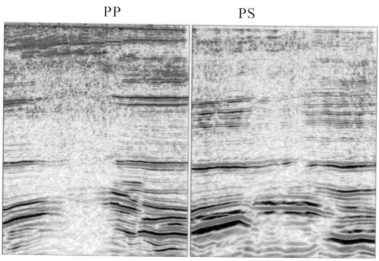

FIGURE 3.31 The PS image (right) reveals details of the reservoir which are hidden on the PP image (left) by effects from the shallow gas. *Courtesy of StatOil ASA. Reprinted with permission.* (For color version of this figure, the reader is referred to the online version of this chapter.)

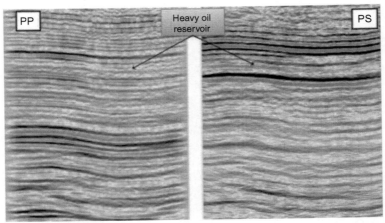

FIGURE 3.32 Transparent reservoir interval on the PP section versus detailed internal architecture on the PS section. *Courtesy of CGG.* (For color version of this figure, the reader is referred to the online version of this chapter.)

Shale volume is a particularly important parameter for the heavy oil recovery processes because shale units act as barriers or baffles to steam movement. Multicomponent data can be used to plan well locations more effectively, leading to a higher recovery factor. Figure 3.32 shows the PP and PS sections from a multicomponent survey in a heavy oil field in Alberta, Canada. The joint interpretation of the PP and PS images provided an improved understanding of the reservoir. While the reservoir interval is almost transparent on the PP section (left), it is clearer on the PS section (on right) with detailed internal reservoir architecture.

3.16 VERTICAL SEISMIC PROFILE

VSP is a technique of seismic measurements used for correlation with surface seismic data. In a VSP, the seismic detectors are placed in a borehole at various depths in the borehole and the source is on the surface. Hydrophones, geophones, or accelerometers are lowered in the borehole and usually clamped at each position. VSP data record the direct arrival from surface source to borehole receivers and also the reflected seismic energy originating from reflectors below the receiver position.

VSPs are the most effective correlation bridge available between the wellbore and the surface seismic data. Since surface seismic data record time delay, not depth, VSP provides correlation of depth versus time measured in seismic. The direct measurement of velocities and time versus depth enables calibration of surface seismic with well information. Using this information, surface seismic data can be more properly imaged and calibrated with well logs. VSP data also provide

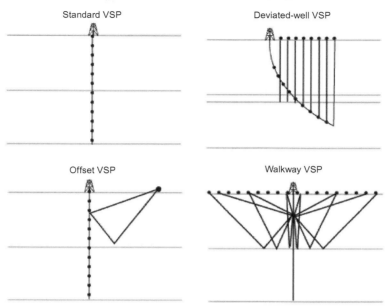

FIGURE 3.33   Various types of VSP acquisition geometries. The standard zero offset VSP is the most common type. The receivers are in the borehole and seismic source on surface. *Graphic ©* *Schlumberger. Used with permission.*

transmission effects between the surface and the reservoir such as average velocity and amplitude loss. VSP data are collected over intervals less than 30 m. Figure 3.33 shows some selected VSP survey geometries.

A zero or near-offset VSP is the standard type (Fig. 3.33) that has the seismic source positioned close to the well head to focus the energy down and ahead of the wellbore. This is the preferred geometry for well correlation. In an offset VSP survey configuration, the energy source is positioned away from the well head. This type of survey provides images at a distance laterally away from the well.

Many extract reflections from VSP data to correlate with those from conventional seismic data. Figure 3.34 shows one such example. In many other cases, the key information obtained from VSP data is direct arrival calibration data and consequently the velocity information. If this is the only reason for the VSP survey, similar information could be obtained by a less expensive approach called check shot surveys. In most instances, sonic logs are only collected over a fraction of the borehole encompassing the reservoir interval. The travel time information from a VSP survey allows the seismic reflection images, which are typically generated as a function of two-way travel time, to be accurately registered with the well log information.

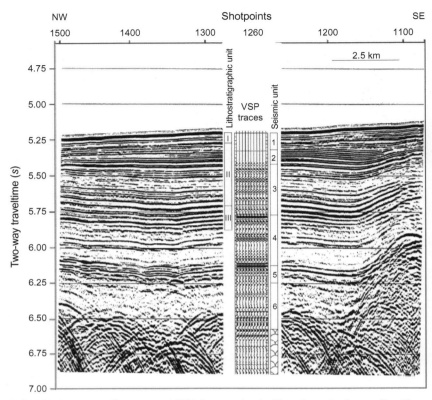

FIGURE 3.34 Zero offset processed VSP data correlated with surface seismic recording. *Courtesy of Texas A&M University.*

3.17 CROSSWELL SEISMIC

The resolution of the seismic data normally is too low to resolve finer subsurface heterogeneities. Uncertainty in velocities can make it difficult for accurate depth positioning subsurface features. Crosswell seismic fills the resolution gap between well logs and surface seismic measurements. Understanding of the continuity and connectivity of reservoirs between wells may be evaluated using cross-borehole seismic. The imaging reveals subtle details of reservoir structure and flow units. The interpretation and integration of the velocity information provided by tomograms and the high vertical resolution reflection images provide useful information. In particular, the crosswell reflection seismic image data provide structural details and lateral resolution on the order of meters that are not seen in surface seismic data.

Figure 3.35 shows an example of a cross well seismic survey. The target reservoir at the Nagaoka site is a water-saturated aquifer at a depth of about 1100 m in the injection well. CO_2 exists in supercritical phase at the depth

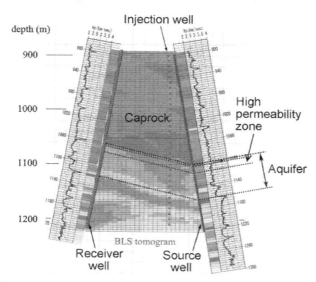

FIGURE 3.35   An example of a cross well survey:- A Cross-section showing the injection and observation wells, sonic logs, and geological structure. The caprock is 130–135 m thick and the aquifer into which CO_2 was injected is 55–60 m thick. From Onishi et al. (2009). (For color version of this figure, the reader is referred to the online version of this chapter.)

and temperature of the aquifer, and its density is lower than that of the formation water. After injection, the CO_2 migrates upward in the aquifer and is trapped below a caprock of mudstone that dips at about 15°.

Figure 3.36 shows crosswell seismic data overlain along with surface seismic data and well logs. Crosswell seismic data improve the understanding of the reservoir geometry, reservoir continuity, and rock properties from the reflection seismogram. The data also provide details of fluid migration, from both the velocity tomography and reflection seismogram.

There are a number of caveats to using crosswell seismic data. First, crosswell is a 2D image of the reservoir and has all the limitations associated with 2D seismic data. Out-of-plane reflections and diffractions can affect data quality and reflection continuity. Second, most vertical wells are drilled down to the reservoir, not well through it. Since crosswell tomography uses recording aperture to recover interwell velocity, tomograms may have unacceptably low resolution in the reservoir interval.

3.18   GRAVITY TECHNIQUES

Gravity surveys can be done by air or on land. The interpretation of gravity data is performed by incorporating seismic and other measurements. The earth's gravity field is affected by the density of different kinds of rocks.

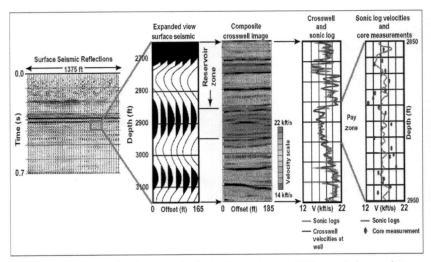

FIGURE 3.36 Crosswell reflection and velocity images (center) fill the resolution gap between 3D surface seismic data (left), and the sonic log data (right center) and core measurements (far right). Crosswell seismic fills the gap *AAPG Explorer*. © *AAPG. Reprinted by permission of AAPG whose permission is required for further use.* (For color version of this figure, the reader is referred to the online version of this chapter.)

Surveys to map these differences can be used by mineral explorers and developers to help locate certain rock formations. Precision instruments known as gravimeters are used to measure the changes in the earth's gravitational field. A gravimeter is designed to measure differences in gravity accelerations rather than absolute magnitudes. In all gravity surveys the vertical component of **g**, g_z, is measured. The gravimeter measures only differences in gravity between two stations. The unit of acceleration to describe measured gravity is *gal*, coined after Galileo. Gravimeters used in geophysical surveys have an accuracy of about 0.01 milligal or mgal (1 milligal = 0.001 centimeter per second per second). That is to say, they are capable of detecting differences in the Earth's gravitational field as small as one part in 100,000,000. Gravity acceleration differences occur because of local density differences. Anomalies of interest are often about 0.2 mgal. Measurements over the survey area are generally made on a grid and the results are mapped and interpreted to reflect the presence of potential oil- or gas-bearing formations. The principle of a gravimeter is the use of a very sensitive spring and weight system attached to a beam. As gravity increases, the weight is forced downward, stretching the spring. The beam is then brought back to a horizontal position; the amount of movement required to do so is proportional to the gravitational force. This information is recorded and later analyzed.

Gravimeters measure the ambient gravitational field at any specific point, or station. From gravity measurements over a specific area, local structural

traps, stratigraphic traps, or fluid movement in reservoir can be inferred if sufficient density contrast exists between the geologic feature of interest and the
surrounding rocks. The differences in gravitational field measurements are
determined by rock mass and depth of burial. The surface-gravity technique
can be applied to any field, depending on reservoir thickness, size, depth of
burial, porosity, and density contrast between the fluids. Rock and fluid density contrasts determine the response. A gravity data survey is generally less
expensive but has less resolving power than seismic exploration survey. Airborne gravity radiometry provides a rapidly acquired, high-resolution image
of local gravity anomalies.

Gravity surveys are generally used in reconnaissance exploration, for
determining large-scale anomalies, as in basin analysis. Given the relatively
low resolution of the gravity data, they need to be augmented by constraining
information. Interpretation of its results is typically not unique. For example,
depth and density in gravity methods are tightly interrelated that to know one
accurately means the other has to be well determined. Expressed another way,
a particular measurement is equally likely to be the result of two different subsurface conditions. A gravity model can be modified in an infinite number of
ways in order to match the observed gravity signal, so a unique consistent
answer needs to be obtained with some other control points. Seismic, well
log, and other geological data provide critical constraining information for
gravity models, limiting the degrees of freedom of our modifications and
enabling us to produce a meaningful, constrained earth model that is consistent with both gravity and seismic datasets. Borehole gravimeters combined
with high-precision surface meters are being used for defining changes in reservoir fluid saturation.

Borehole gravimeters or BHGM can be used to obtain very accurate measurements of the density of horizontally layered rocks. Borehole gravimeters
can detect small changes of porosity in adjacent sediments. This is a modified
version of the surface designed to fit into a sonde, which can be lowered into a
borehole on a wireline. Its precision is on the order of a few microgals. The
instrument is designed to measure the difference in gravity between two stations in a borehole. An important use of borehole gravimeter data is in reservoir evaluation. The tool is used in cased wells to locate gas reservoirs behind
pipe and help identify gas in zones where resistivity logs are problematic,
such as fresh water sands and shaly sands. In addition, the bulk density, which
yields porosity when the pore fluid is known, based on the borehole gravity
data, is the best value obtainable for reservoir engineering use since the instrument investigates a large volume of rock surrounding the borehole. The wide
radius of investigation has also been successfully used to determine gas–oil
and oil–water contacts in reservoirs where other measurements have been
ineffective. BHGM can be used to calculate the difference in oil saturation
between the invaded and undisturbed reservoir zones, which can in turn give
an estimate of movable hydrocarbons.

3.19 MAGNETIC TECHNIQUE

Magnetic and aeromagnetic surveys are passive measurements of the ambient magnetic field. They are used in reconnaissance work, especially for determination of basement features. Additionally, they are used for determining the thickness of sedimentary cover, basement faults and uplifts, basin modeling, structure size, and depth. Precision magnetometers with elaborate electronic instruments are used to measure the variations or anomalies of the ambient magnetic intensity field. Measurements are made either on the ground surface or from low-flying aircrafts. For airborne surveys, tail-stinger-mounted magnetometer on the aircraft is used that samples 10 times a second. A GPS system provides accurate positioning of the magnetometer. The survey airplanes fly at a constant low-flying altitude along closely spaced, parallel flight lines. Line spacing typically is 2–4-km grid at an elevation of about 500 m above the ground. Additional flight lines are flown in the perpendicular direction to assist in data processing. The aeromagnetic surveys are often followed up with surface magnetic surveys that involve stations spaced only 50 m apart. These huge volumes of measurements then are processed into a digital aeromagnetic map. In marine environment, shipboard magnetometers are applied. Figure 3.37 shows examples of both gravity and magnetic airborne survey from Lake Ontario.

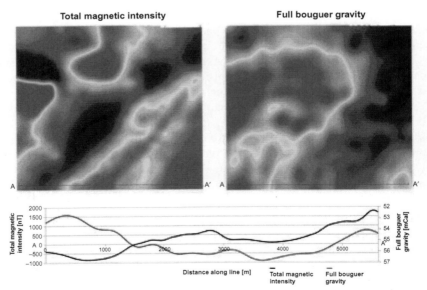

FIGURE 3.37 Gravity and magnetic airborne survey from Lake Ontario. *Courtesy of Sanders geophysics, Ottawa, ONT.* (For color version of this figure, the reader is referred to the online version of this chapter.)

The strength of a magnetic field is measured in units of Gauss (G), or alternatively, in Tesla (T). The magnetic field of the earth at the surface is on the order of 1 Gauss. In the cgs system, when the medium has a permeability of 1 (air or a vacuum), 1 Gauss = 1 Oersted (1 dyne per unit pole). Gauss and Tesla are units of magnetic induction, also known as magnetic flux density.

The intensity of the magnetic field measured on or above the surface of the earth is dependent upon the location of the observation point in the primary magnetic field of the earth and local or regional concentrations of magnetic material. The intensity of the magnetic field on the earth ranges from a minimum of about 0.25 Gauss (G) or 25 microtesla (μT) at the magnetic equator to more than 0.65 G near the magnetic poles. Magnetic anomalies are local variations of the magnetic field produced by magnetic material subsurface layers.

Magnetic anomalies of geologic interest are of two types: induced anomalies and remnant anomalies. Induced anomalies are the result of magnetization of a body by the earth's magnetic field. The anomaly produced is dependent upon the geometry, orientation, and magnetic properties of the body, and the direction and intensity of the earth's field. Remanent anomalies are controlled by the direction and intensity of magnetization and the geometry of the disturbing mass; usually one type of magnetization is dominant and the other can be ignored in the approximation of the results for interpretation. Most magnetic anomalies are a combination of these two types.

3.20 ELECTRICAL AND EM SURVEYS

EM methods have the potential to detect fluid saturation in reservoir. The technique is also used to monitor the movement of fluids within a reservoir. Replacement of brine by gas or oil can cause a change in electrical resistivity of a porous rock of multiple orders of magnitude. Seismic methods, on the other hand, are generally poor at detecting fluid content because the fluid content of a media has much weaker response on changes in acoustic/elastic impedance. Figure 3.38 illustrates the difference between resistivity and velocity properties with fluid saturation. The measured response is a function of rock resistivity, capacitance, and inductance properties.

The electrical resistivity of reservoir rocks is highly sensitive to changes in water and hydrocarbons in pore space of rock formations (from Archie's law). This high sensitivity to water saturation in a reservoir can be exploited by EM techniques where the response is a function of reservoir resistivity (Fig. 3.38). There is a direct correspondence of the change in Sw with the change in the electric field amplitude.

Electrical and EM surveys, originally used mainly in mineral exploration, are adapted to applications in petroleum exploration and reservoir development projects. EM techniques also combine acoustic measurements with EM theory. Maxwell's equations describe how electric charges and electric

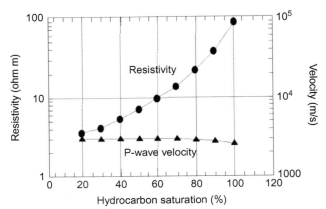

FIGURE 3.38 Seismic P-wave velocity is affected only slightly by hydrocarbon saturation in a porous rock. Resistivity varies more than an order of magnitude (Wilt and Alumbaugh, 1998). *Courtesy: Society of Exploration Geophysicists.*

currents act as sources for the electric and magnetic fields and how they are generated and altered by each other.

Propagating Stoneley waves induce an electric field due to fluid flows through permeable fractures in reservoir layers (Ishido et al., 1981). EM technique is frequently used for mapping subsurface structure and composition by measuring variations in the electrical conductivity of rocks. EM techniques use a grounded electric dipole as a source that is energized with an alternating current at a given frequency to produce time-varying electric and magnetic fields that can be measured on the earth's surface.

The primary field spreads out in space and can penetrate the ground. The depth of penetration is less for higher signal frequencies and for higher ground conductivities; it is therefore important to choose an appropriate frequency for each survey. The primary field will induce a varying voltage in any electrical conductor it encounters. This induced voltage drives another oscillating current in the conductive body, at the same frequency as the primary, but with a phase difference that depends on the electrical properties of the conductor. The secondary current generates another oscillating magnetic field, which can be detected at the surface by a receiving antenna. As the instrument is moved over the survey area, a varying signal will indicate the presence of variations in ground conductivity. EM surveys detect high-resistivity areas often associated with hydrocarbon.

In controlled source EM or CSEM, an oscillating current is generated in a transmitting coil. The source is the flow of controlled pulses of electrical current generated on the ground surface or near the ocean surface. The source penetrates the geological rock layers; the resulting EM is measured using sensors on surface and in boreholes. This indirectly determines the resistivity of the rocks and the types of fluids contained in the rock pores. CSEM has been

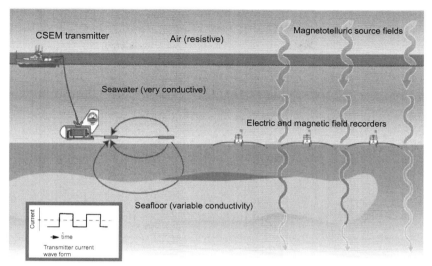

FIGURE 3.39   Marine CSEM data acquisition system. The source is horizontal electric dipole (HED) transmitters. Receivers include an array of electric and magnetic dipole field. The transmitter sends controlled source current continuously. *Image courtesy of Scripps Institution of Oceanography.* (For color version of this figure, the reader is referred to the online version of this chapter.)

called the most significant new technology in oilfield exploration since the development of 3D seismic acquisition. This methodology based on the physics of diffusion of EM fields in the earth rather than wave propagation. This is of a much lower resolution than seismic. The ability to predict reservoir fluid properties ahead of the drill-bit provides a considerable risk reduction for exploration programs. MT data are often acquired with CSEM data, and because these data are sensitive to conductivity, they complement to CSEM.

As shown in Fig. 3.39, a CSEM system uses source with horizontal electric dipole (HED) transmitters and receivers that consist of an array of electric and magnetic dipole field; the source–receivers measure EM impulse response of the earth. In actual field conditions, the transmitter is towed behind a boat, slightly above the seafloor and is continuously moving. The CSEM applications range from shallow groundwater exploration to direct hydrocarbon identification. Interpretation is based on imaging using pseudo-seismic responses as well as 2D and 3D inversion.

However, the most important concept in any EM method is skin depth. EM energy decays exponentially in conductive rocks over a distance given by the skin depth which is proportional to the square root (resistivity/frequency). At a period of 1 s, the skin depth in seawater is about 270 m; this means that over each 270 m the amplitude of EM energy decays another 37%. In 1000 Ωm basalt, at the same period, the skin depth is nearly 16 km, so energy will propagate from the transmitter to the seafloor receivers mostly through seafloor rocks, making the method sensitive mainly to seafloor geology. In comparison, seismic waves decay geometrically as they spread, but retain a resolution that is proportional to wavelength no matter how far they travel. EM signals decay exponentially as conductive rocks absorb energy and have a resolution that is proportional to the depth of the target.

3.21 CROSSWELL EM

Crosswell EM imaging technology is based on radar imaging technology and involves the use of a string of receivers in one well and a transmitter lowered into a neighboring wellbore and moved up and down. Crosswell EM imaging is designed to give accurate measurement of resistivity, hence the oil saturations in the areas between wells. It can provide the engineers with an actual image of fluid migration and show where specific areas of undeveloped reservoir remain. It also has the potential to provide fluid distribution mapping at interwell scale, monitor macroscopic sweep efficiency, reduce uncertainty in the reservoir simulation, optimize the production strategies, plan infill drilling, and thus improve overall recovery.

The technique uses the principles of EM induction and tomography to provide an image of the resistivity distribution between boreholes. Magnetic dipole transmitter loop in the borehole induces currents in the formation; the induced currents are proportional to the transmitter moment (strength), frequency of operation, and formation conductivity surrounding the borehole. Induced currents are inversely proportional to cube of distance from source. Receiver detects direct or primary field and induced or secondary field. Secondary field (formation) is typically 10–50% of the total primary field. The source transmits a continuous sinusoidal signal at programmable frequencies; the signals are detected using an array of induction coil (magnetic field) receivers.

Crosswell EM data are collected from multiple transmitter and receiver positions in the boreholes (Fig. 3.40). They are processed using a two-dimensional nonlinear inversion algorithm to produce a resistivity image of the formation between the wells. The inversion is done using well logs and *a priori* information about the formation to constrain and improve the inverse model. The results provide resistivity image of the section between the wells. This is used for detecting bypassed oil and locating development wells. In a time-lapse application, this can also be used to monitor the movement of a reservoir or of injected fluid.

Figure 3.41 A shows the set up for a Crosswell electromagnetic induction survey. It uses a transmitter tool deployed in one well and a receiver tool deployed in a second well. The tools are connected to specially designed field vehicles. The transmitter broadcasts a frequency that induces currents to flow in underground surrounding rocks. The induced current, in turn, generates a

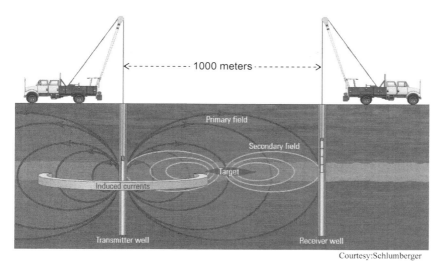

Courtesy:Schlumberger

FIGURE 3.40   Crosswell Electromagnetic survey with dipole EM source and receivers. The EM dipole transmitter is several orders of magnitude greater energy than induction logging tool. An alternating current excites the magnetic dipole transmitter coil to send an electromagnetic field into the formation. This primary field induces eddy currents that, in turn, generate a secondary alternating electromagnetic field whose strength is inversely proportional to formation resistivity. The secondary electromagnetic field is detected at the receiver array along with the primary field. The combination of vertical dipole transmitter and receiver optimizes the survey to map horizontal reservoir layers. Multiple frequencies are used ranging from 5 Hz to 1 kHz. Induced current from source generates secondary field that is influenced by reservoir and detected in receivers in adjacent well. *Graphic © Schlumberger. Used with permission. From Al-Ali et al. 2009.* (For color version of this figure, the reader is referred to the online version of this chapter.)

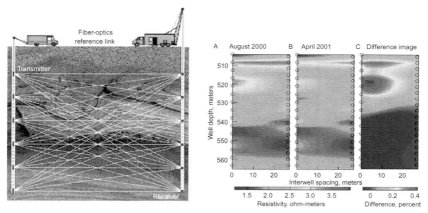

FIGURE 3.41   Crosswell EM data inversion: (a) The survey set up, (b) Monitoring of Resistivity, from LBNL Science and Technology Review, November 2001, https://www.llnl.gov/str/November01/Kirkendall.html. (For color version of this figure, the reader is referred to the online version of this chapter.)

second magnetic field. At the receiver well, a sensor detects the magnetic fields. The receiver is held steady at a fixed depth while the transmitter is lowered over the entire vertical length of the underground zone of interest. Then the receiver is held steady at a fixed depth and the receiver is moved up and down. In this way, researchers create an image of the resistivity of the geologic strata located between the transmitter and receiver.

Figure 3.41 B shows how carbon dioxide flooding can be monitored over time, depicting the two-dimensional images of resistivity in the plane between the two observation wells—one for transmitting, the other for receiving. Image (a) was generated before injection of carbon dioxide and after waterflooding, and image (b) was generated after 3 months of injection. The circles on the left side of each image represent the wells containing the receiver antenna, and the circles on the right side of the images represent the wells containing the transmitting antenna. (c) The difference image is the preinjection image subtracted from the during-injection image and shows areas of change quite clearly. A positive percent difference suggests carbon dioxide is entering the area at the top left. Blue represents water in place, and yellow and red represent the moving oil and carbon dioxide, respectively. Laboratory work is helping to suggest which area is oil and which is carbon dioxide.

In order to obtain a 3D imaging of the reservoir, areal deployment of the receivers is necessary. With the EM source in a well and EM sensors on the surface, borehole to surface survey BSEM has been performed (Marsala et al., 2011). The layout of the survey is shown in Fig. 3.42. The dipole source EM transmitters are deployed in a borehole and the dipole receivers are spread along survey lines on the surface. The receiver array measures amplitude and

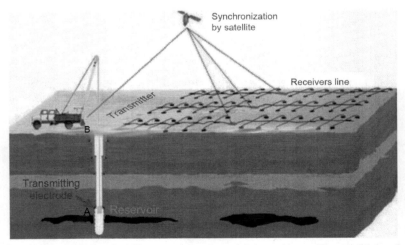

FIGURE 3.42 BSEM method. The transmitting electrode is located in a wellbore, and the receiver array constituting hundreds of electrodes, is placed at the surface. *Courtesy: Saudi Aramco.* (For color version of this figure, the reader is referred to the online version of this chapter.)

the phase of the electric field's radial component, oriented from each receiver station toward the surveyed transmitter well. The coupling of the receivers with the ground and the near-surface heterogeneities are challenges in this technique. In combination with crosswell EM method, BSEM has been successful in mapping in 3D fluid front anomalies and bypassed oil in reservoir. The high-resistivity contrast between oil and injected water provides data that would associate the zones of variable fluid saturation with formation resistivity variations. Using this technique over time lapse would provide changes in the resistivity volume that could be indicative of fluid front movement in the reservoir.

REFERENCES

Al-Ali, Z.A., Al-Buali, M.H., AlRuwaili, S., Ma, S.M., Marsala, A.F., Alumbaugh, D., DePavia, L., Levesque, C., Nalonnil, A., Zhang, P., Hulme, C., Wilt, M., Summer 2009. Oilfield Review 21 (2).

Aminzadeh, F., de Groot, P., 2006. Neural Networks and Soft Computing Techniques, with Applications in the Oil Industry. EAGE Publishing, Amsterdam.

Aminzadeh, F., Brac, J., Kunz, T., 1997. 3D Salt and Overthrust Models. SEG/EAGE 3D Modeling Series, No. 1: Distribution CD of Salt and Overthrust Models, SEG Book Series, Tulsa, Okla.

Bakke, N.E., Ursin, B., 1998. Thin-bed AVO effects. Geophys. Prospect. 6, 571–587.

Batzle, M.L., Hua Han, D., Hofmann, R., 2006. Fluid mobility and frequency-dependent seismic velocity—direct measurements. Geophysics 71 (1), N1–N9.

Bett, M., 2012. The Value of PRM In Enabling High Payback IOR, PESA News Resources, Dec 2012/Jan 2013, p. 45.

Boness, N., 2013. Seismic Data for Reserves Estimation http://www.spe.org/training/courses/SDRE.php.

Castagna, J.P., Swan, H., 1997. Principles of AVO crossplotting. Leading Edge 12, 337–342.

Chopra, S., Marfurt, K.J., 2007. Seismic Attributes for Prospect ID and Reservoir Characterization. Society of Exploration Geophysicists. SEG Publication, Tulsa, Okla.

Clifford, A.C., Aminzadeh, F., 2011. Gas detection from absorption attributes and amplitude versus offset with artificial neural networks in Grand Bay Field [gas detection from absorption and AVO with ANN]. In: Extended Abstracts of 81st SEG Annual Meeting, San Antonio, September 18–23.

Dasgupta, S., Kim, J., AlMousa, A., AlMustafa, H., Aminzadeh, F., Von Lunen, E., 2000. From seismic character and seismic attributes to reservoir properties: case study in Arab-D reservoir of Saudi Arabia. In: 70th Ann. Internl. Mtg: Soc. of Expl. Geophys., pp. 597–599.

Fehler, M., Kelither, P.J., 2011. SEAM Phase 1: Challenge of Subsalt Imaging in Tertiary Basins, with Emphasis on Deepwater Gulf of Mexico. SEG General Book Series, Tulsa.

Haber, E., Oldenburg, D., 1997. Joint inversion: a structural approach. Inverse Problems 13, 63–77.

Hilterman, F.J., 1990. Is AVO the seismic signature of lithology? A case history of Ship Shoal. Leading Edge 9 (6), 15–22.

Kim, K.Y., Wrolstad, K.H., Aminzadeh, F., 1993. Effects of transverse isotropy on P-wave AVO for gas sands: Geophysics. Soc. Expl. Geophys. 58, 883–888.

Loizou, N., Liu, E., Chapman, M., 2008. Petrol. Geosci. 14, 355–368.

Marsala, A.F., Al-Buali, M.H., Ali, Z.A., Ma, S.M., He, Z., Biyan, T., Zhao, G., He, T., 2011. First borehole to surface electromagnetic survey in KSA: reservoir mapping and monitoring at a new scale. In: SPE Annual Technical Conference & Exhibition, Denver Nov 2011.

Mavko, G.M., Nur, A., 1979. Wave attenuation in partially saturated rocks. Geophysics 44 (2), 161–178.

Mavko, G., Mukherji, T., Dvorkin, J., 2003. The Rock Physics Handbook. Cambridge University Press.

Musil, M., Maurer, H.R., Green, A.G., 2003. Discrete tomography and joint inversion for loosely connected or unconnected physical properties: application to crosshole seismic and georadar data sets. Geophys. J. Int. 153, 389–402.

Nissen, S., 2007. Using 3D Seismic Attributes in Reservoir Characterization. 9 August, 2007. Presentation, Hays, KS.

Nolen-Hoeksema, Richard C., 1990. The future role of geophysics in reservoir engineering. Leading Edge 9 (89).

Onishi, K., Ueyama, T., Matsuoka, T., Nobuoka, D., Saito, D., Azuma, H., Xue, Z., 2009. Application of crosswell seismic tomography using difference analysis with data normalization to monitor CO_2 flooding in an aquifer. Int. J. Greenh. Gas Con. 3 (3), 311–321.

Ostrander, W.J., 1984. Plane-wave reflection coefficients for gas sands at non-normal angles of incidence. Geophysics 49, 1637–1648.

Partyka, G., Gridley, J., Lopez, J., 1999. Interpretational applications of spectral decomposition in reservoir characterization. Leading Edge 18 (3), 353.

Robertson, J.D., 1989. Reservoir management using 3D seismic data. J. Pet. Technol. 41 (7), 663–667.

Robison, E., Treitel, S., 2009. Geophysical Signal Analysis. Society of Exploration Geophysicists (SEG) Digital Books.

Rutherford, S.R., Williams, R.H., 1989. Amplitude-versus-offset variations in gas sands. Geophysics 54, 680–688.

Shapiro, S.A., Hubral, P., 1994. A generalized O'Doherty-Anstey ForMula for waves in finely layered media. Geophysics 59, 1750–1762.

Sheriff, R.E., 1980. Monogram for Fresnel zone calculation. Geophysics 45 (5), 968.

Shuey, R.T., 1985. A simplification of Zoeppritz Equations. Geophysics 50, 609–814.

Taner, M.T., 2003. Attributes revisited, http://rocksolidimages.com/pdf/attrib_revisited.htm.

Veeken, P., Rauch-Davies, M., 2006. AVO attribute analysis and seismic reservoir characterization. First Break 24, 41–52.

Veeken, P.C.H. (Ed.), 2007. Seismic reservoir characterisation. In: Handbook of Geophysical Exploration: Seismic Exploration. 37, 355–417 (Chapter 6).

Vozoff, K., Jupp, D.L.B., 1975. Joint inversion of geophysical data. Geophys. J. Roy. Astron. Soc. 42 (3), 977–991.

Wilt, M., Alumbaugh, D., 1998. Electromagnetic methods for development and production: state of the art. Leading Edge 17 (4), 487.

Yilmaz O., 2001, Seismic Data Processing, Investigations in Geophysics, SEG.

Young, R.A., LoPiccolo, R.D., 2003. A comprehensive AVO classification. Leading Edge 22, 1030–1037.

Chapter 4

Formation Evaluation

Chapter Outline

Petrophysical analysis of well logs and cores provide information about formation rocks and fluids in the borehole. Various types of well logs measure different properties in the well. Analysis of the data determines the volume of hydrocarbons present in a reservoir, and its potential to flow through the reservoir rock into the wellbore. This helps us to understand and optimize the producibility of a reservoir.

Developments in Petroleum Science, Vol. 60. http://dx.doi.org/10.1016/B978-0-444-50662-7.00004-4

When oil and gas wells are drilled, physical property measurements are taken using specialized geophysical instrument packages: either on wireline cables after the drill pipe has been removed (wireline logs), or from the borehole while drilling with instruments attached to drill collars (LWD).

4.1   INTRODUCTION

The terms *formation evaluation (FE)*, *well log analysis*, and *petrophysics* are often used interchangeably but do have slightly different meanings. Generally, however, they involve using core and fluid laboratory physical and chemical property measurements and well logs to evaluate wells for potential hydrocarbon reservoir rocks and the volume of economic hydrocarbon accumulations, as well as mechanical properties of the rocks penetrated by a drilling well. The following types of information:

- Laboratory fluid property measurements.
- Laboratory (rock) core physical property measurements.
- Drill cuttings descriptions (strip log).
- Mud logs.
- Wireline log measurements.
- Measurements while drilling (MWD).
- Logging while drilling (LWD).
- Formation flow tests.

Are all used to conduct FE.

FE is used to establish the presence of reservoir rock, evaluate reservoirs for potential hydrocarbons, and estimate the volume of those hydrocarbon reserves. It is also used to develop reservoir mechanical properties models for the drilling and producing departments. Figure 4.1 illustrates the interrelationships between FE and other disciplines within a petroleum exploration and production (*E&P*) operation.

To appreciate the economic role of FE, in the operations of a petroleum company, the formula for calculating original oil (reserves) in place:

$$\text{STOOIP} = \frac{7758 A h \phi (1 - S_w)}{B_{oi}}, \tag{4.1}$$

where STOOIP is stock tank original oil in place, in Bbls; A is the reservoir closure area, in acres; h is the average reservoir (net) thickness, in feet; ϕ is the average reservoir decimal porosity; S_w is the average reservoir decimal water saturation; B_{oi} is the initial oil formation volume factor. 7758 is the acre-ft to Bbls conversion factor (*different conversion constants will be required, for other units*).

A similar relationship also exists for gas reserves.

$$\text{SCFOGIP} = \frac{43,560 A h \phi (1 - S_w)}{B_{gi}}, \tag{4.2}$$

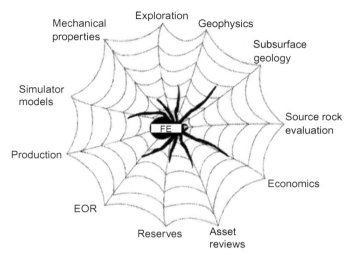

FIGURE 4.1   Formation evaluation (*FE*) relationships in petroleum exploration and production (*E&P*). *Coutesy hillpetro.*

where SCFOGIP is standard (Temperature and Pressure) cubic feet original gas in place; A is the reservoir closure area, in acres; h is the average reservoir (net) thickness, in feet; ϕ is the average reservoir decimal porosity; S_w is the average reservoir decimal water saturation; and B_{gi} is the initial gas formation volume factor. 43,560 is the acre-ft to SCF conversion factor (*different conversion constants will be required, for other units*).

Three of the five variables, in Eqs. (4.1) and (4.2) (h, ϕ, and S_w), are obtained from well logs via FE, or petrophysics. The closure area, A, comes from Seismic Mapping and/or subsurface geology. The Formation Volume Factor, B_{gi}, comes from an analysis of rock and fluid measurements, by a qualified reserves engineer.

In practice, the terms h, ϕ, and $(1 - S_w)$ in Eqs. (4.1) and (4.2) are replaced by:

$$\sum_{i=1}^{n} h_i \phi_i (1 - S_{wi}), \tag{4.3}$$

where h_i can be as small as the sampling interval on the well logs. Obviously, because of the large values possible, for A, small changes in ϕ_i and S_{wi} can result in very large changes in the reserves values. The remainder of this chapter will discuss how a petrophysicist uses FE to arrive at quantitative estimates of reservoir porosities, and saturations, as well as estimates of permeabilities.

4.1.1   Mud Logs

Mud Logs incorporate simplified versions of geologist's sample logs, along with various drilling parameters. They are often used in lieu of geologist's sample

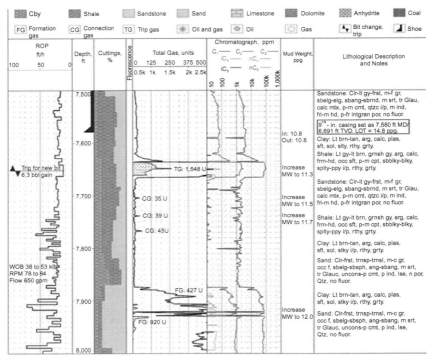

FIGURE 4.2 Mud log example (after Ablard et al. 2012) (© Schlumberger. Reprinted with permission. Schlumberger. Used with permission).

description logs. They also provide an early warning of "Pay Zones," as well as potential "Blow-Outs," "Lost Circulation," and noxious and/or poisonous gases. Drillers utilize them to optimize their drilling operations. Figure 4.2 is an example of a mud log.

4.1.2 Wireline Logs

Wireline logs are records of the physical and/or chemical properties of the materials penetrated by a drilling well. Open-hole wireline logs are acquired by suspending instrument package on a cable to make the measurements and are acquired after the well has been drilled (at least to a casing point), but before casing has been set. They continuously record multiple parameters, including quality control logs. Figure 4.3 shows an open-hole wireline log through a sand-shale (clastic) section, with geologic interpretation.

4.1.3 MWD/LWD Logs

MWD and LWD logs utilize specialized drill collars and data telemetry systems that allow most wireline measurements to be made, as the well is being drilled. Because MWD/LWD systems commonly use mud-pulse telemetry,

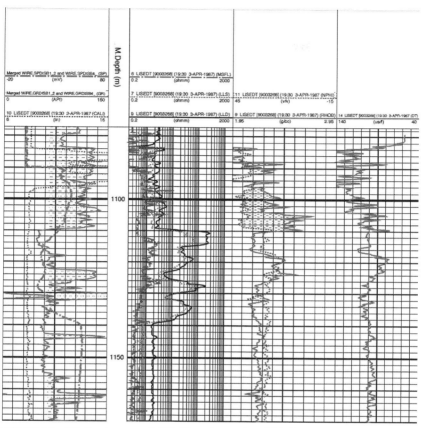

FIGURE 4.3   Wireline log example hill. (For color version of this figure, the reader is referred to the online version of this chapter.)

they are real-time measurements, compared to wireline measurements, which are made only at casing points. They were developed for use in high-risk wells and for high-angle deviated or horizontal wells, which were difficult to log with wireline methods. The LWD/MWD measurements are used during geosteering operations in deviated and horizontal wells.

Figure 4-4 shows the tool set up for LWD / MWD measurements. They are made shortly after the hole is drilled and before the potential complications of drilling or coring operations. This also results in the reduction of fluid invasion compared to wireline logging because of the shorter measurement time. The LWD / MWD tool is battery powered and uses read-only memory chips to store logging data until they are downloaded. The LWD tools take measurements at evenly spaced time intervals and are synchronized with a system monitoring time and depth. After drilling, the LWD tools are retrieved and the data downloaded from each tool.

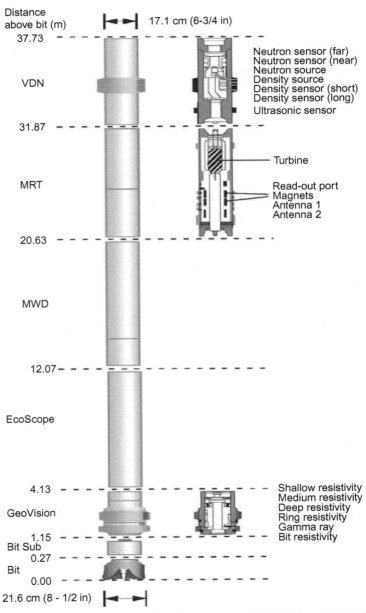

FIGURE 4.4 Configuration of the drill string used for LWD-MWD operations, Sketch by Tom Cattlet, USGS. *source: http://www.netl.doe.gov/technologies/oil-gas/FutureSupply/ MethaneHydrates/rd-program/GOM_JIP/GOM_LWD.html#LWD_Image.*

4.1.4   Who Uses Well Logs?

Petroleum engineers continuously use well logs and petrophysical analyses in the course of their careers.

- Completion engineers utilize petrophysical results to establish completion intervals.
- Reserves engineers utilize petrophysical results to establish recoverable reserves and value to their owners.
- Asset review teams use petrophysical results to establish property values for property disposal and/or acquisition.
- Reservoir engineers utilize petrophysical results to build simulator models and for depletion planning.
- Production engineers utilize petrophysical results to help develop and operate fields.
- Enhanced oil recovery (EOR) engineers utilize petrophysical results to plan EOR operations.

4.1.5   What Is Desired Versus What Is Measured

While three of the five variables, in Eqs. (4.1) and (4.2), are obtained from well logs, two of this three, porosity (ϕ) and saturations (S_w and S_o or S_g), are not measured directly. Instead, well logs measure such things as:

- Acoustic interval transit time, Δt, which depends upon rock type (lithology), porosity (ϕ), saturation (S_w and S_o or S_g), and fluid type, filling the pores.
- Neutron porosity, PHIN (ϕ_N), which depends upon rock type (lithology), porosity (ϕ), saturation (S_w and S_o or S_g), and fluid type, filling the pores.
- Bulk density, RHOB (ρ_B), which depends upon rock type (lithology), porosity (ϕ), saturation (S_w and S_o or S_g), and fluid type, filling the pores.
- Natural gamma radiation (total and/or K-U-T spectral), which depends upon matrix type (lithology).
- Photoelectric factor, PEF, which depends upon matrix type (lithology).
- Spontaneous polarization, SP, which depends upon borehole/formation water salinity contrast and rock type (lithology).
- Apparent resistivity, R, which depends upon ϕ, rock type (lithology), fluid saturation (S_w and S_o or S_g), and formation water salinity.
- Nuclear magnetic resonance (NMR) relaxation times, which depend upon fluid types, pore sizes, and the material lining the pore throats.

The desired porosities and saturations must be inferred from well log measurements, using FE/petrophysical models and techniques.

In spite of the overlapping responses of the various wireline and MWD/LWD tools, most of them are identified with specific applications. Some tools and techniques (e.g., caliper, temperature, and pressure) do have a single primary purpose. Many, however, really have multiple primary purposes. Almost all of them are used as components in various multitool/multitechnique petrophysical analyses.

4.1.6 Uses of Well Logs

Once hydrocarbon indications have been found, these log measurements are used to quantify the reservoir thickness (net pay), pore space (porosity), and the type and amounts of fluids occupying that pore space (water, gas, and oil saturations). This basic reservoir information (net pay, porosity, and saturation) is used with structural and stratigraphic information to develop STOOIP and SCFOG@STIP values for reserves estimates and depletion management (see Eqs. 4.1 and 4.2).

In addition to reservoir volumetric measurements, wireline and MWD measurements also provide information on:

- Borehole volume.
- Mechanical properties of the rocks penetrated by wells.
- Temperatures and pressures of the subsurface.
- Subsurface borehole path.
- Images of the borehole wall.
- Structural information about the subsurface.

4.2 WELL LOGGING TOOLS

4.2.1 Porosity Tools

In the simplest terms, a reservoir rock may be considered to be made up of two components: a solid, matrix, and void space, porosity. Knowledge of porosity is extremely important, for FE, because without porosity, there would be no place to put liquid or gaseous hydrocarbons. Porosity logs are borehole measurements, which respond, primarily, to the porosity (void space) in the rock. Four established borehole measurements—resistivity (conductivity), density, acoustic (sonic) transit time, and hydrogen index (neutron porosity), as well as NMR—all respond to formation porosity.

4.2.2 Resistivity (Conductivity) Tools

Resistivity and induction (conductivity) logs are commonly known as saturation logs because formation resistivity (conductivity) is dependent upon both porosity and saturation. In 1942 and 1950, G.E. Archie, of Shell Oil Co., described the formation factor/porosity and resistivity index/water saturation relationships we now know as Archie's equations:

$$F = \frac{R_o}{R_w} = a\phi^{-m},$$

(4.4)

$$I = \frac{R_t}{R_o} = S_w^{-n},$$

(4.5)

where F is the (dimensionless) formation factor, R_o is the electrical resistivity of a brine-saturated rock, R_w is the electrical resistivity of the brine saturating the rock, ϕ is the (decimal) porosity of the rock, a and m are coefficients determined by the data, I is the (dimensionless) resistivity index, S_w is the (decimal) water saturation of the rock, and n is a coefficient determined by the data.

The empirical coefficient, m, is sometimes called the cementation exponent. The empirical coefficient, a, is sometimes called the tortuosity coefficient. The empirical coefficient, n, is sometimes called the saturation exponent.

Archie's Equations relate formation resistivity to porosity and water saturation, S_w. In the absence of hydrocarbons, $S_w = 1.00$, Archie's first equation can be solved for porosity if the water resistivity, R_w, and Archie coefficients, a and m, are known. For homogeneous reservoirs with water legs, Archie's second equation can be used to evaluate the reservoir qualities of water sands when only resistivity or induction logs are available. To do this, the formation factor in the hydrocarbon leg is assumed to be the same as the formation factor in the water leg. Resistivity (conductivity) logs were the first porosity tools available.

4.2.3   Acoustic (Sonic) Tools

An acoustic impulse will travel through a material at the speed of sound, characteristic of that material. The interval transit time, Δt, is the inverse of the acoustic velocity. Knowledge of Δt is useful not only for estimating formation porosity but also for estimation of seismic velocities for inversion of seismic data. Figure 4.5 is a schematic drawing of one borehole compensated wireline acoustic logging tool. The measurement principle is analogous to a reversed seismic refraction profile. Multiple acoustic receivers, located beyond the critical refraction distance from an acoustic source, record the arrival time of an acoustic impulse, which has been refracted along the borehole wall.

To compensate for the effects of borehole diameter irregularities (rugosity) and sonde inclination, the process is reversed and the arrival times for the sensors are averaged. The interval transit time, Δt, is this average divided by the source-receiver separation.

Different wireline and MWD/LWD vendors offer slight variations of the measurement concept, shown in Fig. 4.5. Modern digital equipment allows specialized sources rich in compressive (longitudinal, or P) and/or flexural (shear or S) wave motion, as well as receiver arrays with automatic semblance picking

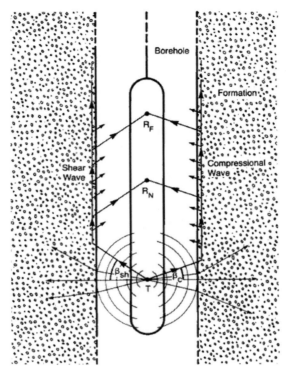

FIGURE 4.5 Compensated acoustic tool. The compressional wave refracts at the critical angle and travels along the borehole wall before detected by the receivers. The shear wave at the left also refracts at shear critical angle and received at the receivers RF and RN. *After Ellis et al. 1987. Courtesy of Schlumberger.*

logic to identify rock P and S wave transit times or inverse velocities. In all cases, however, the basic measurement concept is the refraction time, shown in Fig. 4.5.

Jessie Wyllie and associates at Gulf Research and Development Co. (Wyllie et al., 1958), Louis Raymer and associates at Schlumberger (Raymer et al., 1980), and Jean Raiga-Clemenceau and associates at TOTAL (Raiga-Clemenceau et al., 1988) developed empirical porosity/acoustic transit time relationships for sedimentary rocks. None of these relationships have any physical basis but do appear to fit observed data (some better than others).

Acoustic logs do not seem to be as severely affected by borehole washouts as some of the other porosity tools. For this reason, they are often preferred in rugose borehole situations. Acoustic logs are, however, affected by clay minerals and gaseous hydrocarbons and must be corrected for these effects.

Geophysicists use acoustic logs for computing synthetic seismograms to correlate the geological formations with recorded seismic data. The product of formation velocity from sonic logs and density from density logs is the acoustic impedance. This helps identify the origin of seismic reflections on the recorded seismic data and establish ties between the well log and the seismic section.

4.2.4   Density Tools

The density logging tool is a borehole wall contact tool (see Fig. 4.6). A collimated chemical radioactive γ-ray source bombards the borehole wall with high-energy γ-rays (1.76 MeV) which rattle around, colliding with nuclei and electrons, in the formation. Collisions with nuclei are essentially elastic resulting, primarily, in directional change of the γ-ray trajectories, with very little energy loss.

Because γ-ray energies are attenuated with collisions with any electron, it is essential that the source and detector(s) are placed in close contact with the borehole wall. Density tool is reliable and is used as the primary porosity tool, it is generally used in density–neutron combination tool. Modern density log sondes record not only compton scattering of γ-rays but also those within a lower, *photoelectric effect*, energy window. This latter measurement is useful for lithology identification. γ-Ray attenuation is proportional to electron density, which, for most sedimentary rocks, is roughly proportional to bulk density. MWD/LWD density logs use source-detector sets on four quadrants of the MWD/LWD subassembly to insure that at least one of them is in wall contact, as the drill string is rotated.

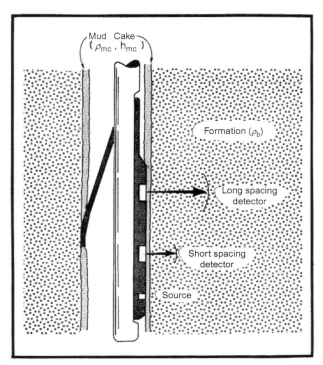

FIGURE 4.6   Compensated density tool. *Courtesy of Schlumberger.*

For a simple reservoir rock, consisting of a solid (matrix) of average density, ρ_{ma}, and porosity, ϕ, filled with a fluid of average density, ρ_f, the bulk density, ρ_B, is given by:

$$\rho_B = \phi \rho_f + (1 - \phi) \rho_{ma}, \qquad (4.6)$$

which can be solved for ϕ as:

$$\phi_D = \frac{(\rho_B - \rho_f)}{(\rho_{ma} - \rho_f)}. \qquad (4.7)$$

Density logs are severely affected by borehole washouts, mudcake buildup, Barite weighted muds, clay minerals, and gas. Density log measurements must be corrected for these effects, before Eq. (4.7) can be used for quantitative ϕ_D estimates. Severe borehole washouts may render density log data unusable.

4.2.5   Neutron Porosity Tools

Neutron logs were the first logging tool developed primarily to measure formation porosity. When fast (>1.0 MeV) neutrons collide with nuclei, within a rock, they are scattered (change trajectory direction) and lose part of their energy to the target nucleus. The amount of the energy loss is related to the relative masses of the neutron and the target nucleus. The greatest (i.e., near total) energy loss occurs when the mass of the target is close to that of the incident neutron (i.e., the hydrogen nucleus).

Within a few microseconds of introduction to a rock, fast neutrons are moderated (i.e., lose energy) to thermal (<0.1 eV) energy levels, and the resulting moderated neutrons drift randomly until they are captured by chlorine, bromine, silicon, or hydrogen atoms. The capturing nucleus then becomes excited and emits a high-energy γ-ray of capture, whose energy is diagnostic of the capturing nucleus.

The hydrogen ion density of a material (often called hydrogen index) can be estimated by counting either the thermal neutron or capture γ-ray flux at some distance from the fast neutron source. Both of these techniques are used in neutron porosity tools.

Figure 4.7 is a schematic rendering of neutron porosity tool measurement principles. The measurement principle is analogous to that of the density tool. A source of fast neutrons is introduced into the borehole wall, where they are scattered and the resulting moderated thermal neutron and/or capture γ-ray flux is monitored at some distance from the neutron source.

Different wireline and MWD/LWD vendors offer slight variations of the measurement concept. MWD/LWD neutron logs use source-detector sets on four quadrants of the MWD/LWD subassembly (i.e., similar to density subassemblies), to insure that at least one of them is in wall contact, as the drill string is rotated. In all cases, however, the basic measurement concept is the neutron flux attenuation, shown.

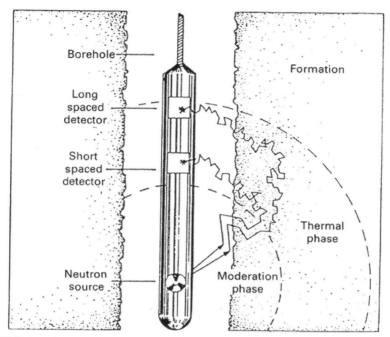

FIGURE 4.7   Compensated neutron tool. *After Ellis (1987).*

Neutron porosity, ϕ_N, is based on the bulk *hydrogen index*, HI_B, which is the ratio of hydrogen ions present in a rock to the number of hydrogen ions present in an equal volume of water. The bulk *hydrogen index* for a simple reservoir rock, consisting of a solid (matrix) of average, HI_{ma}, and porosity, ϕ, filled with a fluid of average HI, HI_f, is given by:

$$HI_B = \phi HI_f + (1 - \phi) HI_{ma}. \tag{4.8}$$

Because of the definition, of HI, its value for water is $HI_w = 1.00$. The HI for most oils is usually assumed to be $HI_o \approx 1.00$. For most reservoir rocks, the average $HI_{ma} \approx 0.00$. With these assumptions and the HI definition, Eq. (4.8) can be simplified to:

$$\phi \cong HI_B. \tag{4.9}$$

The neutron empirical calibration standard is the API limestone test pits, in Houston. For this reason, neutron logs are often presented as limestone porosity, even if the dominant lithology penetrated by a well is not limestone. In these cases, transforms based on models such as that shown in Fig. 4.8 are used to convert the measured ϕ_{Nls} to the appropriate lithology porosity. Cross-plot porosities and lithology analyses are all based on neutron data in limestone units.

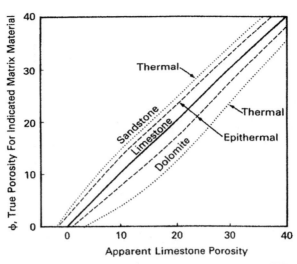

FIGURE 4.8 Neutron porosity tool Lithology Based Porosities. *After Ellis, 1987.*

Neutron logs are severely affected by borehole washouts, mudcake buildup, chlorine, boron, clay minerals, and gas. Neutron log measurements must be corrected for these effects, before quantitative ϕ_D estimates can be made.

Since the introduction of compensated density and neutron tools, the combination of these two porosity tools has become the industry standard for both wireline and MWD/LWD porosity measurement. Not only do they provide excellent porosity measurements, but they also provide information about reservoir rock lithology. In addition, the gas and clay mineral effects on these two measurements work in opposition:

- The effects of clay minerals increase apparent ϕ_N but tend to decrease apparent ϕ_D.
- Conversely, the effects of gas increase apparent ϕ_D but decrease apparent ϕ_N.

Consequently, if both density and neutron logs are available, good estimates of total porosity, ϕ_T, are:

$$\phi_T \cong \phi_D + \phi_N \quad \text{or} \quad \phi_T \cong \sqrt{\frac{(\phi_D^2 + \phi_N^2)}{2}}. \qquad (4.10)$$

4.2.6 Lithology Tools

Essentially, all wireline and MWD/LWD measurements respond to lithological changes in the borehole wall rock. "Lithology tools" are those tools, or combination of tools, which have distinct responses to these changes. Single tool methods utilize a single tool whose lithology response is unique enough that it can be used alone to determine the dominant lithology.

4.2.7 Natural Gamma Ray Logs

There are three natural sources of γ-ray radiation

- Potassium 40 (^{40}K)
- Thorium (^{232}Th) decay series
- Uranium (^{238}U) decay series

Of these three sources, ^{40}K is the major contributor to rock natural γ-radiation because potassium is a major constituent of feldspars, sylvite, micas, and clay minerals.

Micas and clay minerals are major constituents in shales and contain large amounts of potassium. Because potassium is not a constituent of quartz, calcite, or dolomite, the natural γ-ray log is often used as a shale indicator. Uranium and thorium salts are often precipitated in reducing environments, such as organic shales (e.g., the Antrim Shale, of the Michigan Basin). A low γ-ray "Sand Line" (100% sand) and a high γ-ray "Shale Line" (100% shale) are established for the interval of interest, and intermediate γ-ray values are assigned shale volume, V_{sh}, by a variety of empirical γ-ray V_{sh} models. Similar V_{sh} estimates can be made for carbonate reservoir rocks.

The γ-ray log is not an infallible V_{sh} indicator. Arkosic sands contain large amounts of orthoclase and microcline (K-Feldspars), which will generate high γ-ray responses. Zircon and Sphene, often found in high-permeability, clean, beach sand (good quality reservoir rock) deposits, commonly have thorium as contaminants which will generate high γ-ray log responses. Monazite sands are often mined as a source of thorium, and carbonates often have uranium and thorium contamination along fractures, which will also generate high γ-ray log responses. Sylvite and other potash minerals contain large amounts of potassium, which can generate high γ-ray log responses in evaporite sequences.

The γ-ray log works as a "Shale indicator" and is commonly used to estimate V_{sh}.

4.2.8 Spontaneous Potential

Spontaneous potential (SP) is an electrochemical phenomenon that occurs when the salinity of the drilling mud filtrate is different from that of the formation waters *and* high-permeability rocks are bounded by low-permeability shales (which form cation-selective membranes). When the formation waters and mud filtrate are not the same, salinity anions (negatively charged ions) and cations (positively charged ions) will migrate from the higher-concentration fluid to the lower-concentration fluid, in an attempt to equalize the salinities.

Shales act as cation-selective membranes, passing only cations and blocking anions, leaving a cation deficiency on the upstream (high permeability) and a cation surplus on the downstream (low permeability) borehole surfaces. The resulting electrical charge variations can be monitored by measuring the electrical potential, with respect to some fixed point, as a function of depth.

A "normal" SP occurs when the formation waters are more saline than the mud filtrate waters. In these cases, the SP opposite sands will be negative, compared to that opposite shales. A "reversed" SP occurs for the opposite situation.

Because the SP opposite (high permeability) sands is different from that opposite (low permeability) shales, the SP is often used as a sand-shale indicator. An SP "Sand Line" (100%s and) and a "Shale Line" (100% shale) are established for the interval of interest, and the intermediate SP values are scaled linearly between 0% and 100% V_{sh}.

4.2.9 Photoelectric Factor

Low-energy (<0.2 MeV) γ-rays are adsorbed by electrons, within the rock, increasing their energy level and releasing a photoelectron, on impact. This process is called *photoelectric adsorption* and is the principal upon which the Z, or PEF, curve of modern density log tools is based.

Photoelectric absorption is dependent upon the atomic number (Z) of the target atom. Lithodensity, spectral density, or Z-logs use the near detector ratio of γ-ray flux in the compton scattering and photoelectric absorption windows to estimate bulk PEF, which is largely independent of porosity. Because of this, the PEF curve can be an excellent lithology indicator.

The accepted matrix PEF values for common reservoir dominant minerals are:

- Calcite: 5.1
- Dolomite: 3.1
- Quartz: 1.8

which provides sufficient separation for lithology identification, in complex lithology environments, such as the Pinda Formation of the Congo Basin.

Less well known is that the accepted PEF values for common potassium feldspars are:

- Anorthoclase: 3.1
- Microcline: 2.8
- Orthoclase: 2.9

which are all around the dolomite value. The reason this is significant is that Arkosic sands, which have high potassium feldspar content, seldom occur in carbonate environments. As a result, high γ-ray responses, in a clastic (sand and shale) environment, such as California's San Joaquin Valley, with PEF values near 3.0 indicate Arkosic sands, not shale or dolomite.

The PEF values for micas and clay minerals, the principal components of shales, range from 1.8 to 6.3, which mean that the PEF alone is not a good clay and mica mineral identifier. Because the PEF curves are based on low-energy γ-rays, they are very sensitive to Barite mud additives and cannot be used for lithology identification in heavy mud situations.

4.2.10 Saturation Tools

Saturation tools are those logging tools, which are sensitive to (gas, oil, and water) saturation variations. Formation resistivity, R_t, estimated from these tools is used to estimate the uninvaded formation water saturation, S_w. The hydrocarbon saturation, S_o or S_g, is then $1 - S_w$. There are four types of open-hole saturation tools:

- Resistivity tools which use electrodes (e.g., laterologs).
- Resistivity tools which use coils (e.g., induction logs).
- Resistivity tools which use antennas (e.g., dielectric or electronic propagation logs).
- NMR logs.

While three of these tools measure formation resistivity, they each interact with the formation differently.

4.2.11 Electrode Tools

Electrode tools utilize Galvanic coupling between the sonde and the formation, so they require conductive fluid in the borehole. *Electrode tools should not be used in wells drilled with air, foam, mist, or OBM.*

Formation resistivity was the first borehole measurement that the Schlumberger brothers made. The Pechelbronn well was logged by dropping Schlumberger surface resistivity electrode array down the borehole and making a series of measurements at different depths similar to what had been previously been done, on the surface, with a technique called resistivity profiling.

Doll (1951) introduced the laterolog design, which removed the large metal electrodes required for guarded electrode measurements. The major purposes of the successive electrode designs were twofold:

- Enhance thin-bed resolution and quantitative measurement.
- Decrease the influence of borehole fluids, mudcake, and invasion.

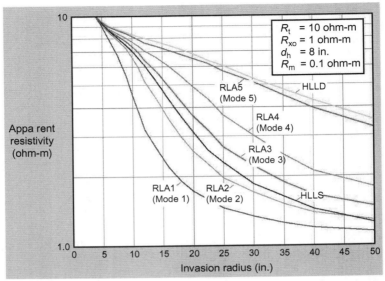

FIGURE 4.9 Comparison of various High-Resolution Laterolog Array Tool® (HRLA) apparent resistivities versus invasion, for an 8 in borehole containing conductive mud. *Courtesy of Schlumberger.* (For color version of this figure, the reader is referred to the online version of this chapter.)

These two goals tend to be mutually exclusive. The initial response to this is to use multiple electrode or coil arrays on the same sonde.

The electrode tools of choice are array laterolog tools. The Schlumberger version of this tool, introduced in 1998, uses downhole microprocessors, multiple electrode arrays, all current electrodes on the sonde, multiple frequencies, and records both amplitudes (impedance), and phase shifts to provide six apparent resistivities, which can be inverted to provide two-dimensional (cylindrical) resistivity models (Anon, 2000).

Figure 4.9 shows the invasion effects for a Schlumberger HRLA® tool in an 8-in. borehole filled with conductive mud. Laterolog type tools work best when the borehole fluid resistivity, R_m, is less than that of the formation, R_t. They are calibrated for 8 in. boreholes filled with nominal (e.g., 15,000–25,000 ppm NaCl) salinity mud. Significant departures from these conditions require borehole, or environmental corrections, before quantitative water saturation, S_w, estimates can be made.

4.2.12 Microelectrode Tools

Microelectrode tools are contact devices, designed to provide flushed zone resistivities, R_{xo}. In permeable formations, there is usually significant flushing or fluid invasion of formation fluids by the mud filtrate. This flushing, or invasion, is observable by separation of the micro- and deep resistivity

Laterologs should not be used in air, mist, or oil based mud filled boreholes.

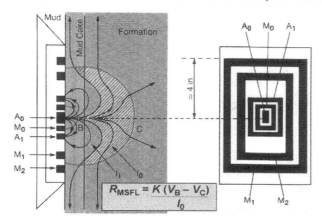

FIGURE 4.10 Microspherically focused log (MSFL) array and operational schematic. *Courtesy of Schlumberger.* (For color version of this figure, the reader is referred to the online version of this chapter.)

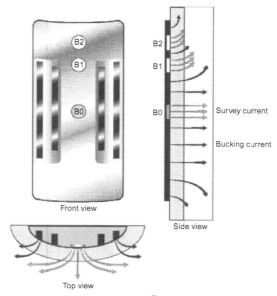

FIGURE 4.11 MicroCylindrically Focused Tool® (MCFL) Array and operational schematic. *After Ellis and Singer (2007). Courtesy of Schlumberger.*

curves and provides a qualitative indication of permeability. The resulting flushed zone water saturation, S_{xo}, is estimated from R_{xo}. The movable hydrocarbons (i.e., hydrocarbons that can be produced) are $S_{xo} - S_w$ (Figs. 4.10 and 4.11).

The current microresistivity tools of choice are either the microspherically focused log or the microcylindrically focused tool. Microelectrode tools are very sensitive to mudcake thickness and must be corrected for it, if accurate flushed zone resistivities are desired.

4.2.13 Induction Resistivity Coil Tools

Coil, or induction, resistivity tools were developed to be able to make formation resistivity measurements in air, mist, foam, or OBM filled boreholes (Doll, 1949b). They can also be used with freshwater mud filled holes and modest formation resistivities, such as in the U.S. Gulf of Mexico and Niger Delta. Induction tools really measure formation conductivity (the inverse of resistivity) and analog induction tool resistivity values over 200 ohm-m (<5 mmho/m conductivity) are highly suspect. Modern digital system induction tools, with downhole A/D conversion and microprocessors, can accommodate higher formation resistivities but are still not as reliable as laterolog tools, for the highest formation resistivities. Induction tools should not be used for situations with saline muds and high-resistivity formations.

Figure 4.12 shows the operational principals of induction logging. A high-frequency (kHz) oscillating current is passed through a transmitter coil, which generates a primary oscillating magnetic field. This primary magnetic field generates secondary oscillating currents in the formation surrounding the borehole (and in the borehole fluids), which generate secondary oscillating magnetic fields.

Micro-electrode tools should not used in air, mist, or oil-based mud filled boreholes.

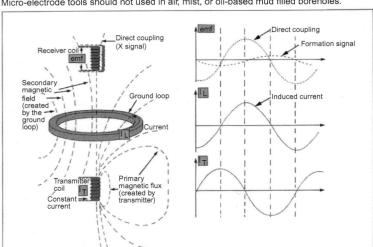

FIGURE 4.12 Induction log operational schematic. *Courtesy of Schlumberger.* (For color version of this figure, the reader is referred to the online version of this chapter.)

A receiver coil, coaxial with, but distanced from, the transmitting coil detects the combined primary and secondary magnetic fields and feeds this signal to software which separates the primary and secondary signals and inverts them to apparent formation resistivity (conductivity). Digital induction tools, with downhole A/D conversion and microprocessors, can measure both impedance and phase, providing more precise formation models.

The induction tools of choice are array induction tools, first introduced by BPB Oilfield Services, now Weatherford (Martin, 1984).

The original Schlumberger AIT® (array induction tool) contained a single transmitter coil system operating simultaneously at three frequencies, with in-phase and quadrature signals measured at six of eight receiver coil arrays. This downhole sonde delivered 28 conductivity measurements, which were borehole corrected, deconvolved, and combined to form sets of five depth of investigation log curves for sets of "vertical resolution" "log" resistivity. The five resulting log curves do not need environmental corrections and represent the effective resistivities of five concentric resistivity model rings in a borehole-centered cylindrical model. The deepest (R90) log curve is often taken as R_t and the shallowest (R10) is often taken as R_{xo}. The on-board computer, however, can also deliver one final resistivity model inversion to estimate R_t and R_{xo}, which may or may not be significantly different from R90 and R10, respectively (Figs. 4.13 and 4.14).

4.2.14   Antenna (Dielectric) Tools

Antenna, or dielectric, resistivity tools were developed for use in situations with heavy, viscous, oils and very freshwaters. Water is one of the very few naturally occurring electrically polar molecules. As a result, freshwater has a very high ($\sim$80) dielectric constant. By contrast, oils and most reservoir rocks have dielectric constants less than 9 (i.e., a factor of approximately 10, or more, less).

At one time, all major logging vendors and some operating companies had versions of dielectric tools. The various tools fell into the two distinct styles, shown in Fig. 4.15. Low (MHz) mandrel type tools had coils attached to nonconducting sondes and were centered in the borehole (LHS of Fig. 4.15). High (GHz) pad-type tools used slot antennas, pressed up against the borehole wall (RHS of Fig. 4.15).

In late 2010 and early 2011, Schlumberger and Halliburton introduced new generation digital dielectric tools. The market for these new tools, however, appears to remain heavy oils with freshwaters.

4.2.15   NMR Tools

The utility of measuring Nuclear Magnetic Resonance or NMR in reservoir rocks was recognized in the 1950s (Brown and Fatt, 1956), and a prototype

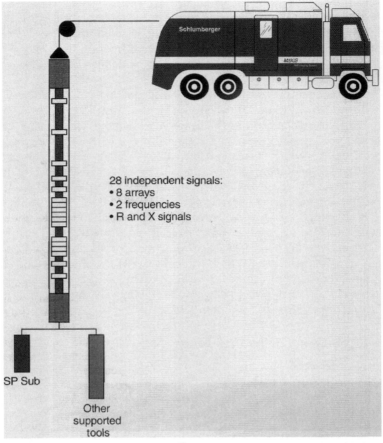

28 independent signals:
- 8 arrays
- 2 frequencies
- R and X signals

SP Sub

Other
supported
tools

FIGURE 4.13 Array Induction Imager Tool® (AIT) schematic. *Courtesy of Schlumberger.* (For color version of this figure, the reader is referred to the online version of this chapter.)

tool was developed in the same decade (Brown and Gamson, 1960; Hull and Coolidge, 1960). A reliable NMR logging tool was not developed, however, until the late 1980s (Coates et al., 1991). It took 30 years for the technology required for a reliable logging tool to catch up with the promise provided by scientific insight and laboratory measurements. Magnetic resonance is a phenomenon by which a nucleus of an atom absorbs electromagnetic radiation of a specific frequency in the presence of strong magnetic field. NMR tools are used in detecting light atoms Hydrogen in hydrocarbons. The NMR logging tool consists of permanent magnets that project a magnetic field into the formation. The magnetic field align the proton spin axis of the reservoir fluid. The lines up the north pole of the nuclei of the fluid with the south pole of

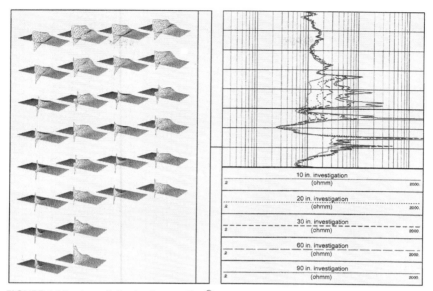

FIGURE 4.14   Array Induction Imager Tool® (AIT) signals and ring model resistivity curves. *Courtesy of Schlumberger.* (For color version of this figure, the reader is referred to the online version of this chapter.)

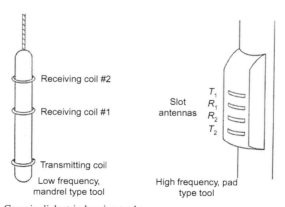

FIGURE 4.15   Generic dielectric logging tools.

of the magnet in the tool. The measured time for this alignment is T1 and is related to the viscosity of the hydrocarbons. Another magnetic field that is oscillating is then applied (using RF transmitter-receiver) in perpendicular or transverse direction to the first set and in resonance with the nuclear motion. This tips the nucleus away from the direction of the permanent magnets in the NMR. This makes the nucleus to precess or to go into an orbital motion. At that point using RF signals, a series of evenly time spaced magnetic field pulses are applied in reverse direction of the permanent magnets. The precession of nucleus created by these successive pulses is allowed to return in the original field direction of the permanent field. This relaxation time T2 measured in milliseconds is the precession decay time. For permeable rocks with hydrocarbons the decay time is longer. T1 defines distribution of pore sizes in the reservoir rocks. T2 distribution is used for predicting total porosity, bound-fluid porosity, permeability and pore throat sizes (Freedman 2006).

The ion with the greatest NMR response is the hydrogen ($^1H^+$) ion, with no neutrons or electrons. Oxygen and carbon, by contrast, have only paired protons and a very low NMR response. Consequently, the primary source of NMR in reservoir rocks is due to the hydrogen ions in reservoir rocks, or due to the hydrogen ions in water and hydrocarbon molecules.

In the absence of any external magnetic field, proton spins have random orientations. In the presence of an external magnetic field, the proton spin moment will orient either parallel or antiparallel to the external field. If a larger (local) magnetic field is imposed, the proton spins will reorient, with their moments precessing (rotating) around the new net magnetic field, much like the gyroscope inertial field processing around the earth's gravitational field. This proton spin precession about the net external magnetic field vector is called *proton precession*, or (in the presence of a pulsed local field) *NMR*. The frequency of the rotation is proportional to the magnetic field strength. The length of time for the precession to decay is related to the energy interaction between the spinning protons and the material in the walls of the vessel containing water and/or hydrocarbons. The precession decay time, for vessels of similar composition, is thus the size of the vessels (i.e., pore sizes, for reservoir rocks) and the composition of the vessel wall material lining (i.e., the pore throat lining).

NMR logging tool (Fig. 4.16). The original application of NMR logs was to estimate porosity and saturations in freshwater, heavy oil reservoirs. A second application was to estimate reservoir rock pore size distributions, and thereby permeability. Figure 4.17 shows a comparison of laboratory core and NMR permeability analyses. Figure 4.18 shows a comparison of wireline NMR and laboratory core porosity and permeability analyses for high-resolution (HR) CMR® tools. The conventional gamma ray (GR) and density logs indicates poor reservoir development while CMR indicates good reservoir quality which is corroborated with core data. CMR tool defines tight interbedded layers in the reservoir internal. The conventional GR, density logs, however, do not show any tight interbed in the reservoir zone.

Coates et al. (1999) listed the following NMR log applications:

- Presence and quantities of different fluids (water, oil, and gas).
- Porosity and pore size distributions.

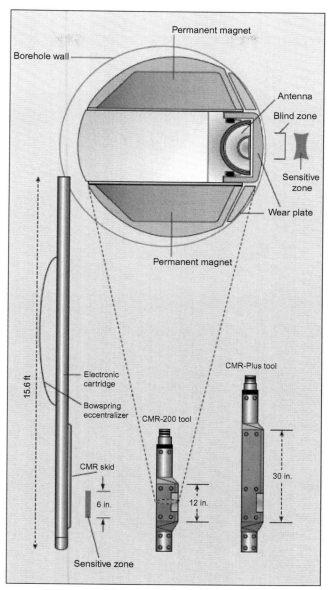

FIGURE 4.16　Modern NMR logging tool. (© *Schlumberger. Used with permission*). (For color version of this figure, the reader is referred to the online version of this chapter.)

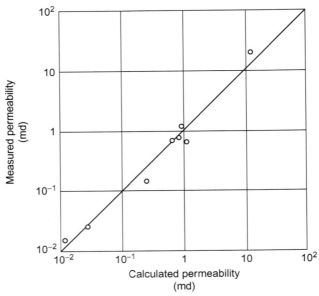

FIGURE 4.17 Comparison of measured and NMR relaxation time predicted core permeabilities. *After Seevers (1966). Reprinted with permission of SPWLA.*

- Bound and free water saturations.
- Effective porosity and permeability.
- Flushed zone saturation, S_{xo}, for wells drilled with OBM.

Figure 4.19 shows a CMR® analysis, showing wireline porosity, permeability, saturation, and fluid type analyses.

NMR is independent of lithology. While the NMR response to reservoir rock matrix material is not as strong as for the traditional porosity (acoustic, density, and neutron) tools, the precession decay time, as indicated earlier, is related not only to the pore sizes but also to the material lining the pore throats. To get the most reliable NMR log interpretations, an independent source of reservoir lithology is required. Kenyon (1997) provides a very good summary of the petrophysical principles behind NMR log applications, including the lithology aspects.

NMR logs are less sensitive to matrix lithology than conventional (acoustic, density, and neutron) tools. However, they do require a good independent source of lithology information, for best results. They also provide permeability estimates, which definitely fill a gap in the conventional porosity log tool kit. NMR logs, however, are more expensive to run than conventional porosity tools, in terms of logging charges, rig time, and post processing. For best NMR log analysis results, conventional porosity tools such as density and neutron logs are required to be run.

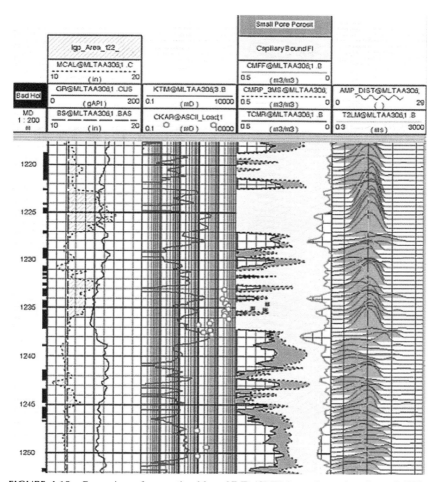

FIGURE 4.18   Comparison of conventional logs, NMR (CMR) log and core data (interval 1232–1236M). The results from CMR data show that bound fluid occupies almost the entire pore space in the rock. Since the relaxation time of heavy oil is so close to bound water that the CMR log derived permeability, the average is about 73 md, which is much lower than core permeability (Tangyan et al., 2005). (For color version of this figure, the reader is referred to the online version of this chapter.)

4.2.16   NMR Logging Saturation

NMR tools can discriminate between different fluid types (see Fig. 4.20) even though their depth of investigations is very shallow. For best results, however, these NMR interpretations require supporting information from porosity, lithology, and resistivity tools.

4.2.17   Salinity (R_w) Tools

Archie's equations require porosity (from porosity tools), resistivity (from resistivity saturation tools), and formation water resistivity. Salinity tools

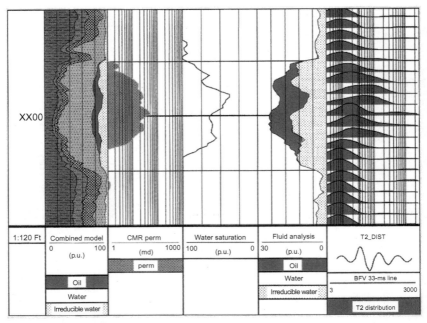

FIGURE 4.19 CMR® interpretation, showing porosity, permeability, saturation, and fluid type analyses. *Courtesy of Schlumberger.* (For color version of this figure, the reader is referred to the online version of this chapter.)

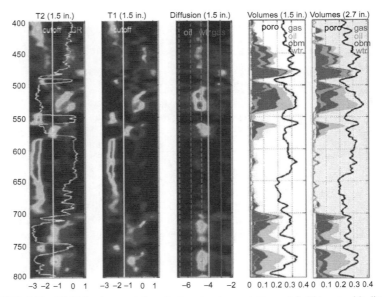

FIGURE 4.20 NMR log interpretation display showing variations of fluid types with distance into the formation. *After Minh et al. (2007). AAPG © 2007. Reprinted by permission of AAPG whose permission is required for further use.* (For color version of this figure, the reader is referred to the online version of this chapter.)

provide the last element. If a formation water sample is available, its resistivity can be measured directly. Sometimes (particularly) older wet chemistry analyses do not include direct R_w measurements. In those cases, equivalent NaCl salinities and R_w can be estimated from the individual ionic and TDS concentrations (Arps, 1953; Moore et al., 1966).

The effects of cation-selective membranes in low-permeability shales, adjacent to high-permeability reservoir rocks, were described earlier in the *lithology tools* section. The potential differences created can be used to estimate formation water R_w via the Nernst equation (Wyllie, 1949):

$$E = -\frac{RT}{F} Ln \sqrt{\frac{a_w}{a_{mf}}} \tag{4.11}$$

where E is the relative (i.e., from clay to sand) SP anomaly, in mv; R is the universal gas constant; F is the Faraday constant; T is the absolute temperature, in $°K$; a_w and a_{mf} are the formation water and mud filtrate electrochemical activities, respectively.

A strong SP will develop opposite clean sands, bounded by clays, if the formation water salinity is significantly different than the mud filtrate salinity. In those cases, R_w can be estimated from the mud filtrate resistivity, R_{mf}, and the SP. Weak R_w/R_{mf} contrasts yield anemic SP deflections. The presence of clay minerals in the sand will also depress the SP deflection.

SP-based R_w estimates tend to be lower than other estimates. The SP, however, can be used to estimate R_w in oil and gas sands, with no water leg; something that other log-based R_w estimators cannot do.

In clean water sands, Archie's equations can be solved for R_w, as $S_w = 1$. The resulting R_w estimate is called R_{wa}, or *apparent water resistivity*. This technique is extremely fast and may be one of the most commonly used means of estimating R_w. This technique will work only with environmentally corrected deep resistivity data and requires knowledge of the Archie a and m coefficients. It will not work, however, in the presence of clay minerals and/or hydrocarbons. It also requires the assumption that the same brine be present in the hydrocarbon zone as the water zone.

In clean water sands, the ratio of the deep to microelectrode resistivities can also be used to estimate R_w. This technique does not require knowledge of Archie model coefficients, like the R_{wa} technique. It does, however, require environmentally corrected deep and microelectrode resistivities and is subject to all of the other R_{wa} limitations.

4.2.18 Borehole Imaging Tools

Borehole imaging tools, or borehole imagers, provide detailed (mm–cm scale) images of the borehole walls.

Borehole imagers provided whole well mm–cm scale images of the borehole wall, which could be virtually manipulated and examined much like whole cores, but without the time and expense involved in obtaining whole cores.

The two most successful borehole imagers have been:

- Acoustic borehole televiewers (BHTVs).
- Microresistivity scanners.

Each of these imagers has its own advantages and limitations. The real game changer for borehole imagers has been the image workstations. These are powerful microcomputer-based systems which allow the interpreter to manipulate and measure features on the borehole images, much like one would do with physical whole cores, and generate tables of dip and strike information to present alongside other log data.

4.2.19   Acoustic Borehole Imagers

BHTVs are really SONAR devices that emit ultrasonic (500 KHz–2 MHz) acoustic pulses into the mud column and measure the *reflected* amplitudes and (time-of-flight) reflection times. High resolution acoustic imaging device (the CBIL) by Western Atlas use a rotating transducer to fire an acoustic pulse at the borehole wall, with the amplitude and transit time of the returned signal being used to construct high resolution circumferential images. Circumferential Acoustic Scanning Tool® (CAST) from Halliburton is a borehole imaging tool. Figure 4.21 shows CAST tool specifications and an example of CAST pseudocore image.

Acoustic borehole televiewers require liquid-filled boreholes but operate better in OBM than do microresistivity scanners, provide full 360° image coverage of the borehole walls, which microresistivity scanners do not, and operate better in rugose boreholes than do microresistivity scanners. The time-of-flight data also allow three-dimensional information about the borehole wall. Figure 4.22 shows four rotated pseudocore images of the same depth interval, from a Congo Basin well. The protrusions are to mud filled vugs in the borehole wall.

4.2.20   Microresistivity Scanners

Microresistivity Scanners were developed to overcome some of the shortcomings of early BHTV tools. Multiple button electrode arrays were added to dipmeter pads. The problem introduced, with this approach, is that, as the borehole diameter increases, gaps in circumferential coverage between the dipmeter pads expand. The dilemma faced by logging vendors was how to distribute this loss in circumferential coverage to best describe the stratigraphy and structural patterns displayed on the borehole wall. Schlumberger uses four pads, Baker-Atlas and Halliburton use six, (Fig. 4.24) and Weatherford uses eight.

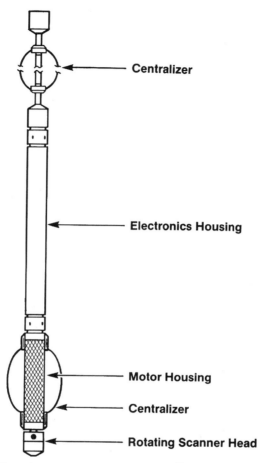

FIGURE 4.21   Circumferential Acoustic scanning Tool® (CAST). *After Goetz et al, 1990. Courtesy of SPWLA.* (For color version of this figure, the reader is referred to the online version of this chapter.)

The larger the number of pads, the smaller the circumferential coverage gap, between each pad, but also the smaller the circumferential area covered by each pad. The smaller the number of pads, the larger the circumferential coverage on each pad, but also the larger the gap in coverage between pads.

Schlumberger, with only four pads, added side flaps to each of its pads to increase the circumferential coverage of each pad the pad flap may not have the same pressure against the borehole wall as the main pad, particularly with continued usage. Because all pads must be in good contact with the borehole wall, this is a serious concern. Figure 4.23 shows example of resolution comparison between electrical resistivity based image logs and acoustic borehole imager or

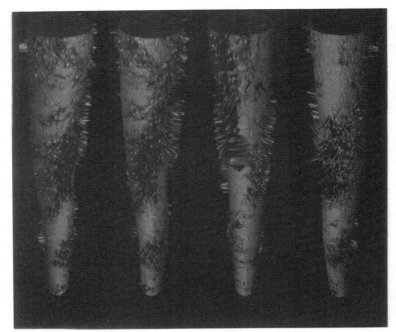

FIGURE 4.22 Rotated acoustic borehole image. Pseudo-Core Images of the Same Depth Interval, from a Congo Basin Well, Showing Mud Invaded Vugs in the Borehole Wall. *Courtesy of www.hillpetro.com.* (For color version of this figure, the reader is referred to the online version of this chapter.)

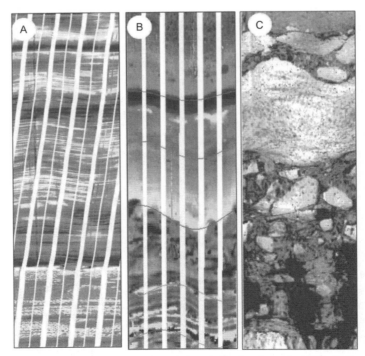

FIGURE 4.23 Examples of modern high resolution wireline imaging tools. (a) The FMI resistivity imager (Schlumberger). (b) The STAR resistivity imager (Baker Atlas). (c). The CBIL acoustic imager (Baker Atlas). *Courtesy of Petroleum Exploration Society Australia.* (For color version of this figure, the reader is referred to the online version of this chapter.)

FIGURE 4.24   Halliburton Electrical MicroImaging® (EMI) tool. *After Anon (2011). Courtesy of Halliburton.* (For color version of this figure, the reader is referred to the online version of this chapter.)

televiewer. Schlumberger use the formation micro imager or FMI tool. All major wireline vendors provide resistivity based imaging service. Wireline imaging the Schlumberger FMI® image. Baker Hughes STAR high resolution resistivity formation image and Baker Hughes CBIL or circumferential borehole imaging log to analyze structural dip, fracture system and depositional environment.

Baker-Atlas and Halliburton Figure 4.24, with six pads, and Weatherford, with eight pads, staggered their pads into two slightly displaced rows of out-sized pads, to maximize circumferential coverage on each pad and minimize the gaps between pad coverage (Fig. 4.25).

Both imaging tools offer only images and neither delivers quantitative measurements. Acoustic imaging tools were developed, primarily, for fault and fracture identification. Microresistivity scanners were developed, primarily, for detailed stratigraphic evaluation. The microresistivity scanners offer millimeter scale image resolution, while the acoustic images offer only centimeter scale resolution. The acoustic imagers work better in OBM filled boreholes than do the microresistivity scanners and offer full 360° borehole wall coverage, which the microresistivity scanners do not. Acoustic imagers also offer three-dimensional depth resolution, which the microresistivity scanners do not Microresistivity scanners may work better in horizontal boreholes,

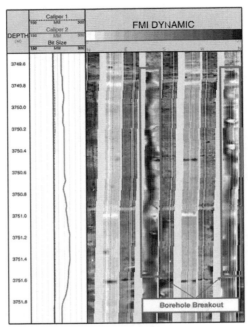

FIGURE 4.25 From Iran Ilam carbonate. Formation breakout observed on an FMI log in well B. The breakout is identified as a pair of poorly resolved conductive zones observed on opposite sides of the borehole (outlined in bold) and showing caliper enlargement in the same direction (note caliper 2 greater than caliper 1). The breakout pictured herein is oriented approximately NNW -SSE and thus indicates a present-day S_{Hmax} orientation of approximately ENE -WSW. *From Rajabi et al. 2010. Reprinted with permission.* (For color version of this figure, the reader is referred to the online version of this chapter.)

because of the need for the acoustic imagers to be centered in the borehole. Acoustic imagers must be run at slower (1200 ft/hour) logging speeds than microresistivity scanners. Each tool has its niche market and the tools are not interchangeable.

4.2.21   Pulsed Neutron Capture Geochemical Logs

Neutron interactions with, and capture by, target nuclei release γ-rays, which are unique to each interaction. Figure 4.26 shows example of pulse neutron with core analysis results from the lab. The results indicate a good correlation in identifying rock mineralogy in the rocks. The resulting interaction γ-ray flux and energies is utilized for matrix and fluid identification (Caldwell, 1958; Martin, 1956). Early pulsed neutron γ-ray spectroscopy included chlorine identification (Stroud and Schaller, 1958), Si/Al ratios (Wichmann and Webb, 1969), and carbon/oxygen ratios (Tittman and Nelligan, 1959).

High-resolution solid-state scintillation counters, allow more precise estimation of elemental and, consequently, mineralogical composition of the material surrounding a borehole.

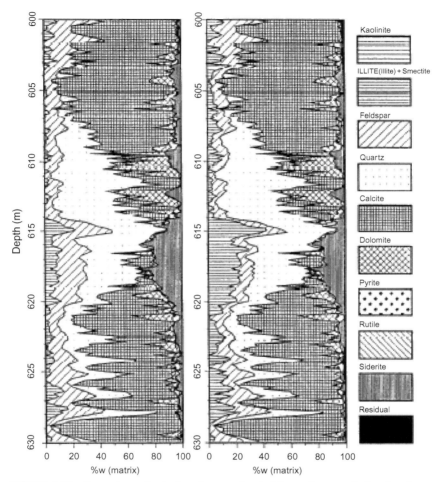

FIGURE 4.26 Comparison of pulsed neutron geochemical log and core analysis mineralogy. *After Van den Oord (1990). Reprinted with permission of SPWLA.*

REFERENCES

Ablard, P., Bell, C., et al. 2012. The expanding role of mud logging. Schlumberger Oilfield Review Spring 2012.

Anon, 2000. High Resolution Laterolog Array Tool. Schlumberger Wireline and Testing, Houston.

Anon, 2011. EMI® Electrical Micro Imaging Service. Halliburton Wireline and TestingPerforating, Houston.

Archie, G.E., 1942. The electrical resistivity log as an aid in determining some reservoir characteristics. Trans. AIME 146, 54–62.

Archie, G.E., 1950. Introduction to petrophysics or reservoir rocks. Bull. AAPG 34 (5), 943–961.

Arps, J.J., 1953. The effect of temperature on the density and electrical resistivity of sodium chloride solutions. Trans. AIME 198, 327–330.

Brown, R.J.S., Fatt, I., 1956. Measurements of fractional wettability of oilfield blocks by nuclear magnetic relaxation method. AIME Pet. Trans. 207, 262–264.

Brown, R.J.S., Gamson, B.W., 1960. Nuclear magnetism logging. AIME Pet. Trans. 219, 201–209.

Caldwell, R.L., 1958. Using nuclear methods in oil-well logging. Nucleonics 16 (12), 58–66.

Coates, G.R., Xiao, L., Prammer, M.G., 1999. NMR Logging Principles and Applications. Halliburton Energy Services, Houston.

Doll, H.G., 1949b. Introduction to induction logging and application to logging of wells drilled with oil base mud. Pet. Trans. AIME 186, 148–162.

Doll, H.G., 1951. The Laterolog: a new resistivity logging method with electrodes using an automatic focusing system. Trans. AIME 192, 305–316.

Ellis, D.V., 1987. Well Logging for Earth Scientists. Elsevier, New York.

Ellis, D.V., Singer, J.M., 2007. Well Logging for Earth Scientists. Springer, Dordrecht, The Netherlands, 1987.

Freedman, R., 2006. Advances in NMR logging, distinguished author series. JPT 58 (1), 60–66.

Goetz, J. F., Seiler, D. D., Edmiston, C. E., "Geological and Borehole Features Described by the Circumferential Acoustic Scanning Tool" SPWLA Transactions, Vol. 31, paper 90-C.

Hull, P., Coolidge, J.E., 1960. Field examples of nuclear magnetism logging. J. Petrol. Technol. 12 (8), 14–22.

Kenyon, W.E., 1997. Petrophysical principles of applications of NMR logging. Log Analyst 48 (2), 21–43.

Martin, P.M., 1956. Well Logging with an Atom Smasher, Petroleum Branch, AIME, Paper 722-G.

Martin, D.W., Spencer, M.C., Patel, H., 1984. Digital Induction—A New Approach to Improving the Response of Induction Measurement, Trans., SPWLA, vol. 25, Paper M.

Moore, E.J., Szasz, S.E., Whitney, B.F., 1966. Determining formation water resistivity from chemical analysis. Trans. AIME 237, 273–276.

Minh, C., Weinheber, P., Wichers, W., Gisolf, A., Caroli, E., Jaffuel, F., Poirier, Y., Baldini, D., Sitta, M., Tealdi, L., 2007. Using the continuous NMR fluid properties scan to optimize sampling with wireline formation testers. In: Expanded Abstract AAPG International Conference and Exhibition, Cape Town, South Africa, October 26–29, 2008.

Raiga-Clemenceau, J., Martin, J.P., Nicoletis, S., 1988. The concept of acoustic formation factor for more accurate porosity determination from sonic transit time data. Log Analyst 29, 54–60.

Rajabi, et al., 2010. Subsurface fracture analysis and determination of in-situ stress direction using FMI logs: an example from the Santonian carbonates (Ilam Formation) in the Abadan Plain, Iran. Tectonophysics 492, 192–200.

Raymer, L.L., Hunt, E.R., Gardner, J.S., 1980. An Improved Sonic Transit Time -to-Porosity Transform, 21st Annual Logging Symposium, SPWLA, Paper P.

Seevers, D.O., 1966. A Nuclear Magnetic Method of Determining the Permeability of Sandstones, SPWLA Transactions, vol. 7, Paper L.

Stroud, S.G., Schaller, H.E., 1958. New Radiation Log for the Determination of Reservoir Salinity, SPE of AIME, Paper No. 1118.G.

Tangyan, L., Zaitian, M., Junxiao, W., Hongzhi, L., 2005. Integrating MDT, NMR log and conventional logs for one-well evaluation. J. Petrol. Sci. Eng. 46 (1–2), 73–80. http://dx.doi.org/10.1016/j.petrol.2004.09.001.

Tittman, J., Nelligon, W.B., 1959,1986. Laboratory Studies of a Pulsed Neutron Source Technique in Well Logging, Society of Petroleum Engineers Paper 1227-G.

Van den Oord, 1990. Experience with Geochemical Logging, 31st Annual Logging Symposium, SPWLA, Paper T.

Wichmann, P.A., Webb, R.W., 1969. Neutron Activation Logging for Si to Al Ratios, Society of Petroleum Engineers Paper 2550.

Wyllie, M.R.J., 1949. A quantitative analysis of the electrochemical component of the S.P. curve. J. Petrol. Technol. 1 (1), 17–26.

Wyllie, M.R.J., Gregory, A.R., Gardner, G.H.F., 1958. An experimental investigation of factors affecting elastic wave velocities in porous media. Geophysics 23, 459–493.

Chapter 5

Geostatistics and Other Unconventional Statistical Methods

5.1 OVERVIEW

In most oil exploration and production problems, we are dealing with limited and incomplete data. We are constantly trying to extrapolate information from sparse measurements (e.g., limited well data and core data on the one hand and large volumes of seismic data with limited spatial resolution). We resort to statistical methods to accomplish this. Among conventional statistical methods (CM) used are: regression analysis, clustering, cross-plotting, principal component analysis (PCA), and spatial statistics/geostatistics. More recently, some unconventional statistical methods such as fuzzy logic (FL), neural networks (NNs), and fractal methods have been found to be useful as well.

In this section, we will give a brief description of fundamentals of conventional geostatistics and unconventional methods. For more details, see Deutsch and Journel (1998), Caers (2003), and Aminzadeh and de Groot (2006).

5.2 CONVENTIONAL GEOSTATISTICAL METHODS

Traditional statistical methods for both spatial and temporal extrapolation have been uses in E&P for several decades. One of the main uses of statistics has been for reservoir characterization through integrating information and data from various sources with varying degrees of uncertainty with different scales and data resolution such as log and seismic data. Other applications include establishing relationships between measurements and reservoir properties; reserve estimation and oil field economics with the associated risk factors. The following are some of the topics to be discussed in this section: matrix plot, correlation, regression, PCA, variogram, kriging, and clustering.

Matrix plot and correlation: In many cases, we need to understand the relationship between different parameters in the reservoir. Often times, significant insight could be derived from "cross-plotting" many parameters against each other. Figure 5.1 shows an example of "matrix plot" where different parameters: net sand, porosity, Vshale, and RMS amplitude are plotted against each other.

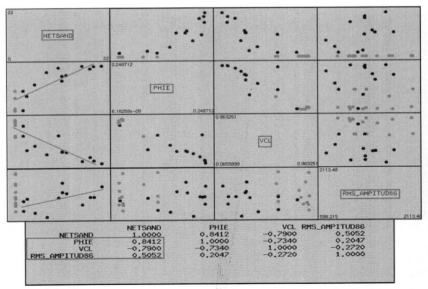

FIGURE 5.1 A matrix plot of net sand, porosity, Vshale, and RMS amplitude. (For color version of this figure, the reader is referred to the online version of this chapter.)

The points in each cross-plot show the ensemble of respective parameters obtained from different wells at the horizon of interest. The RMS amplitude values are the obtained from the seismic sections at the corresponding well and depth location. The matrix at the bottom shows the correlation (R) between different parameters. The closer the correlation number to 1, the better correlation exists between those parameters. A negative correlation indicates an increase in the value of one parameter will be associated with a decrease in the value of the other parameter. Usually, the sign independent square of correlation coefficient (R^2) is used for various comparisons (Dasgupta et al. 2000). A higher R^2 would imply a stronger correlation (independent from whether it is a positive or negative correlation). In the above example, there is a positive and relatively strong correlation between the value of net sand thickness and the porosity (the box under the top left box). Conversely, there is a negative correlation between the sand thickness and Vshale.

Linear regression analysis attempts to establish a linear relationship between a parameter and one or more other parameters. It is a statistical technique for estimating the relationships among variables. In Fig. 5.1, the straight line fitted to the points (on far left boxes) shows the best linear relationship between the value of net sand against porosity, Vshale and RMS amplitudes. Regression is also done to express a quantity in terms of many other variables. As an example, Kaiser and Yu (2012), using regression analysis, express the value of the market capitalization (CAP) of different oil and gas companies against their gas reserves (Rboe), reserves to production ratio (R/P), and debt–equity ratio (D/E) using the relation in Eq. (5.1):

$$CAP = a + b \, Rboe + c \, R/P + d \, D/E \qquad (5.1)$$

Figure 5.2 shows a good correlation between CAP and reserves for several major oil companies (2010 data). In this figure, the correlation is strong (R^2 value of 0.75). The correlation becomes stronger when the other two parameters (R/P and D/E) are included in the regression analysis (R^2 value of 0.83).

5.2.1 Factor Analysis and Principal Component Analysis

One requirement for the regression analysis to work effectively is that the variables in the regression need to be uncorrelated with each other. To ensure that is the case, a process of factor analysis (FA) or PCA is used. *PCA* uses an orthogonal transformation to convert a number of correlated parameters into a set of uncorrelated variables called *principal components*. PCA usually yields a smaller number of new variables than the original ones with the extra benefit of "dimensionality reduction." PCA is defined in such a way that the first principal components have the largest possible variances or "eigenvalues" (i.e., accounts for as much of the variability in the data as possible). Since PCA is sensitive to the relative scaling of the original variables, usually it is proceeded by a data transformation. PCA is closely related to FA. FA typically

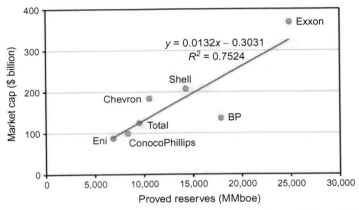

FIGURE 5.2 The correlation between company market capitalization and reserves. Kaiser and Yu, 2012. *Courtesy: Oil & Gas Financial Journal.* (For color version of this figure, the reader is referred to the online version of this chapter.)

incorporates more domain-specific assumptions about the underlying structure and solves eigenvectors of a slightly different matrix. Aminzadeh and Chatterjee (1984) used FA for clustering of different seismic attributes resulting in better prediction of "bright spots."

Data transformation: In many situations, we may need to perform data transformation, for example, the scale of some of the reservoir properties such as permeability and span a large range. When performing PCA, correlation, or other statistical analysis such as clustering or variogram calculation (to be discussed later), appropriate transformation would prevent a highly skewed data distributions. In the case of permeability, a common type of data transformation is to create its logarithm:

$$y = \log(k)$$

Other types of data transformation are to scale the data so that the largest and smallest values are within a given range (e.g., between 0 and 1 or −1 and). After such transformations, one can perform all statistical analyses on the transformed data and transform the results back to the original values at the end.

Variogram An important element of geostatistics is variogram or correlogram. It gives a measure of the spatial variability of data. The variogram for lag distance h is defined as the average squared difference of values separated approximately by h. Equation (5.2) gives the analytical expression for variogram value, $\gamma(h)$:

$$\gamma(h) = \frac{1}{2N(h)} \sum_{N(h)} [z(u) - z(u+h)]^2 \tag{5.2}$$

where $N(h)$ is the number of pairs for lag h. Figure 5.3 shows a typical variogram, with its three main components, nugget, sill, and range.

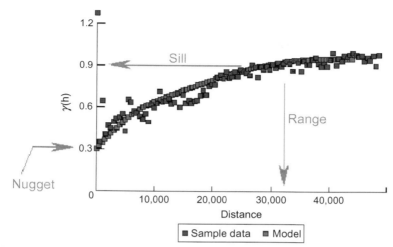

FIGURE 5.3 A typical variogram. *From http://fs.fed.us/ne/fia/gis/workshop/4def_desc.html.* (For color version of this figure, the reader is referred to the online version of this chapter.)

The following are brief descriptions of nugget, sill, and range:

Nugget n: The height of the jump of the variogram at the origin.
Sill s: Limit of the variogram tending to infinity lag distances.
Range r: The distance in which the difference of the variogram from the sill becomes negligible. In models with a fixed sill, it is the distance at which this is first reached; for models with an asymptotic sill, it is conventionally taken to be the distance when the variance first reaches 95% of the sill.

A variogram or correlogram is a single view of the data. We would get a somewhat different picture of the data if we change the lag spacing and/or the range of distances. Also, strong univariate characteristics of the data, such as the large number of 0 porosity and thus highly skewed distribution, can mask some of the spatial structure that is really in the data. This is another reason for transforming the data, such as normal-scoring, can reveal a variogram/correlogram with substantially more visible spatial structure. In general, "it is easy to mask spatial continuity by a poor choice of lag spacing, direction angles, or a poor handling of outlier values. It is rare to generate spatial continuity that does not exist." (Deutsch and Journel, 1998, pp. 58–59).

After examination of the data for the need for data transformation PCA or coordinate transformation (to be consistent with the survey area or reservoir model) we choose the number of directions and the actual directions to create variograms. Often times, we choose two perpendicular directions and one (45°) or two (30° and 60°) azimuthal directions number of lags (*N*) and the lag distance (*h*) are chosen based on the type and scale of the data. We choose

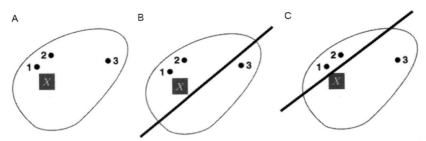

FIGURE 5.4 Three existing well location 1, 2, 3 with a new well at x: (a) no fault plane, (b) fault plane to the south of well x, and (c) fault plane to the north of well x. (For color version of this figure, the reader is referred to the online version of this chapter.)

the lag distance in conjunction with the data spacing. N and h are also dependent of the maximum distance (from a field level to the core scale as will be discussed in this chapter).

Kriging is the weighted averages of the sample data values—taking the distance, direction, and redundancy of neighboring points into account using that model defined from the variogram. It is designed to be the best linear unbiased estimate. To illustrate how kriging can help in a spatial extrapolation of the parameters, we pose the following question. In Fig. 5.4a, points 1, 2, and 3 represent three wells with a reservoir with known reservoir thickness at a given target horizon. With no additional information, what is the best estimate for the reservoir thickness at location X if the values at locations 1, 2, and 3 are 30, 20, and 70 ft and the distance from X to 1, 2, and 3 are 100, 200, and 700 ft?

The answer can be obtained by using the simple "inverse distance" approach given by Eq. (5.3):

$$Z^*(x) = \sum_{i=1}^{n} \lambda_i Z(x_i) \tag{5.3}$$

where weight of λ_i is proportionate to the inverse of distance between the well x and the wells at locations 1, 2, and 3 derived from:

$\lambda_1 = (1/100)/(1/100 + 1/200 + 1/700) = 1/(1 + 1/2 + 1/7) = 0.61,$
$\lambda_2 = (1/200) \times 61 = 0.305,$
$\lambda_3 = (1/700) \times 61 = 0.087$

Thus, the thickness value at the X location is

$$Z(x) = (30 \times 0.61 + 20 \times 0.305 + 70 \times 0.087) = 30.5 ft$$

The above assumes no other information such as geologic data (represented by fault in Fig. 5.4b and c) or seismic data. The fault location in Fig. 5.4b would indicate closer similarity to the thickness values of wells 1 and 2, while the fault location at Fig. 5.4c, would imply more similarity with well 3 and less influence of wells 1 and 2 independent from the proximity of well x to those wells. This is because the reservoir properties for wells on the same side of the fault blocks are more likely to be similar.

In general, aside from distance and geologic faults we should also consider data equalization which may imply giving zero weights to some data points to avoid redundancy or having spatial data saturation or having too many data points at given locations. Preferential direction for changes in the data values in certain direction (anisotropy) is another factor. For example, if we have independent information on thickening or thinning of the reservoir in certain direction, somehow that information should be incorporated in the weights. The same is true if we know of specific patchiness or continuity of the reservoir (e.g., certain stratigraphic information). Much of these factors would be captured by the variograms per the earlier discussion. The following describes a methodology to incorporate all these factors using a case history to be discussed further in this chapter.

5.2.2 Stochastic (Monte Carlo) Simulation

Monte Carlo simulation is a type of stochastic simulation that performs risk analysis by building models of possible results by substituting a range of values—a probability distribution—for any factor that has inherent uncertainty. It then calculates results over and over, each time using a different set of random values from the probability functions. Depending upon the number of uncertainties and the ranges specified for them, a Monte Carlo simulation could involve thousands or tens of thousands of recalculations before it is complete. Monte Carlo simulation produces distributions of possible outcome values. If we do not know the exact distribution of a random variable we want to estimate, we can take samples from that distribution and average them. If we take enough samples, then the "law of large numbers" says this average must be close to the true value. The central limit theorem says that the average has a Gaussian distribution around the true value.

For example, we may want to measure the area of a reservoir unit with a complicated, irregular outline. The Monte Carlo approach is to draw a square around the shape and measure the square. If we through darts into the square, as uniformly as possible. The fraction of darts falling on the shape gives the ratio of the area of the shape to the area of the square. We can cast almost any integral problem, or any averaging problem, into this form. We need to have good criteria to tell if we are inside the outline, and we need a good way to figure out how many darts we should throw. We also need a good way to throw darts uniformly, that is, a good random number generator.

We do not strictly need to sample independently. We can have dependence, so long as we end up visiting each point just as many times as we would with independent samples. This is useful, since it gives a way to exploit properties of Markov Chain in designing the sampling strategy, and even of speeding up the convergence of the estimates to the true averages.

5.2.3 Conditional Simulation and Sequential Gaussian Conditional Simulation

Unlike kriging, instead of creating a single best estimate, conditional simulation generates many, equally probable, alternative realizations. From this set of estimates, an entire distribution function can be built for each cell, representing the range of possible values. See Fig. 5.6 for a practical example of generating different simulations.

Using the model of the variogram/correlogram calculated from the normal-scored data, sequential Gaussian conditional simulation (SGCS) first transforms the data into a Gaussian distribution. Subsequently, SGCS selects one grid node at random and performs kriging of the value at that location. Next step is drawing a random number from a Gaussian distribution that has been constructed to have a variance equivalent to the kriged variance and a mean equivalent to that kriged value. The random value chosen from that distribution is the simulated value for that grid node. Then, other grid nodes are selected randomly and the process is repeated, including all previously simulated nodes in the kriging calculation. This preserves the spatial variability as modeled in the variogram. When all nodes are simulated for an individual realization, it back transforms the values to the original distribution. This gives us the first realization. It then repeats for all the other realizations starting with a different initial grid node using random number generator.

Cokriging, through creating a "cross-variogram" of say well data and seismic data, allows for additional information to be incorporated in the estimates. The cross-variogram is then used to create a cokriged map. Figure 5.5 shows the cokriged porosity map using the cross-variogram of the seismic and well data.

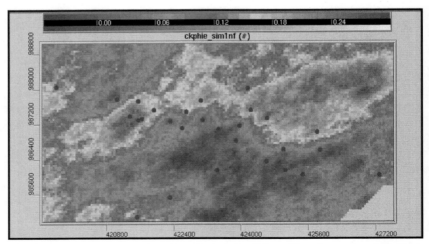

FIGURE 5.5   Cokriged of porosity using seismic and well data. (For color version of this figure, the reader is referred to the online version of this chapter.)

It should be emphasized that there are many alternatives for performing kriging and cokriging. Among these methods are: kriging with external drift, collocated cokriging, and full cokriging of well data and seismic using the cross-variogram of the RMS amplitude and net sand values as well as variograms of each data component. For more details on these methods and their benefits, see Deutsch and Journel (1998).

In what follows, we will provide a workflow work using different above mentioned geostatistical concepts to carry out a typical geostatistics-based "reservoir characterization" more details to follow in Chapter 6 where the focus is indeed the static reservoir characterization.

5.2.4   Workflow-Flow Diagram

To accomplish the task of geostatistical reservoir characterization (e.g., generating the porosity and thickness maps), we describe a conventional approach. Table 5.1 shows the workflow of the process and the use of different methods that to generate reservoir property models.

TABLE 5.1 Step-by-Step Description of the Methodology

Step	Details and Comments
Data preparation	Data editing, choice of boundaries for mapping, and grid size
Data examination	Creation of histograms and matrix plots
Attribute choice	Compare seismic attributes and well properties correlations
Spatial statistics	Calculate variogram and correlogram
Kriging	Extrapolate well properties away from the well using well property variogram models
Kriging with external drift	Extrapolate well properties away from the well using variogram models and seismic at grid points as a guide
Collocated kriging or cokriging	The same as kriging except the seismic information usage is not limited to grid points only
Cokriging	Extrapolate well properties away from the well using variogram and cross-variogram models and seismic data
Simulation	Create multiple (100) realizations of cokriged results
Risk analysis and interpretation	Based on the simulation results, create the predicted value and associate uncertainty at the proposed well location
Ground truth test	Test the prediction results against new wells and examine to ranges of predicted values and true drilling results

Figure 5.6 shows the flow diagram with the specifics geostatistical method employed in a case history in Lobo Field, Venezuela, shows implementation of the workflow described in Table 5.1. Specifically, the porosity (PHI) as one of the well data information is *cokriged* (this and other highlighted geostatistical concepts were discussed earlier) with seismic amplitude information (INST AMP), using the *variogram* models and *cross-plot* of PHI and INST AMP. Fifty *simulated* models of porosity are generated, leading to the creation of the porosity field in the reservoir horizon of interest (bottom right display). P_{20} values of PHIE (better than 80% chance for the actual porosity will be more than the value calculates: it will be further discussed in the reserves part of Chapter 6) appeared to be most consistent with well data and geologically. It was used to improve the interwell porosity distribution in the channel facies area.

Data preparation and examination are very important to do quality control and to get a general feel for the available data and to understand the ranges of different parameters. Examination of data is facilitated by the use of different statistical tools such as histograms, cross-plots, matrix plots, and cross-correlation discussed earlier and shown in Fig. 5.1.

All these methods can be applied to create any reservoir property. The confidence level on mapping results, however, is dependent on the well coverage and extent of correlation between the well properties and seismic attributes. Figure 5.5 showing porosity map generated from cokriging of porosity against RMS amplitude is only one of the many realizations. But, it did demonstrate the value of data integration through cokriging. The net sand map showed a clear E–W channel-shaped sand body that could not be identified

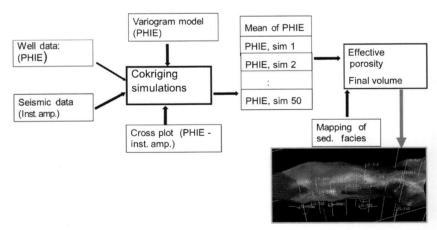

FIGURE 5.6 Flow diagram of reservoir characterization process. (For color version of this figure, the reader is referred to the online version of this chapter.)

previously using well data or seismic data alone. As we will see in Chapter 6, additional simulations (to the right of Fig. 6.6) would allow us to create an ensemble of different realizations providing us with the confidence level (or error bars) associated with the average sand distribution map. This would also allow us to create various limits of the reserves estimates, also to be discussed in more detail in Chapter 6.

5.3   UNCONVENTIONAL STATISTICAL METHODS

Over the past two decades, a new brand of statistical methods, referred to as soft computing (SC) have found their way into many practical applications including the petroleum arena. Where conventional statistical means are deemed inadequate to tackle practical problems, we can employ nontraditional SC methods such as artificial neural networks (ANNs), FL, and genetic algorithms (GAs). These methods and their applications in many petroleum engineering and exploration problems have been discussed in details by many, including Wong et al. (2002), Mohaghegh (2000), and Nikravesh et al. (2003). In this section, we provide a brief overview of each. We also show how these methods can be combined with each other and with the CM to benefit from the strength of each.

5.4   ARTIFICIAL NEURAL NETWORKS

ANN is a nonlinear optimization and information processing tool that attempts to mimic some of the features of the human brain both during training and problem-solving stages. ANN is most suitable for pattern matching, classification, clustering, and approximation through "learning" or "training." In contrast, conventional statistical algorithms and computing techniques are better for precise computation such as fast search algorithms, linear programming, arithmetic calculations, and implementing partial differential equations.

An artificial neuron, also known as perceptron, has similar elements as those of a biological neuron. The key component here is "the processing element," comprised of the integration and activation elements (Fig. 5.7). The integration operator (usually linear) creates a weighted sum of all the inputs. The weights on each "connection" or "link," show the significance and impact of the input parameter on the output. In an ANN weights are determined during the training or learning stage, which typically starts from a randomized "initial state." This is followed by the activation function which is usually a nonlinear operator (e.g., a sigmoid function shown in Fig. 5.7). Equation (5.4) shows the analytical relationship or the transfer function between the input parameters and the output:

$$y = f[S(w_i x_i)] \qquad (5.4)$$

where f in this case is the signum or sgn function, also known as a hard limiter or threshold function. The output of signum or sgn function is 1 or -1, depending upon whether the input to this function is positive or negative.

ANNs are comprised of a large number of neurons similar to the one shown in Fig. 5.7. They are configured in different fashions depending upon the type of the ANN. The simplest one depicted in Fig. 5.8 is called a feed-forward ANN comprised of the input layer (different production data in a well), a number of "hidden layers and an output layer" (in this case the predicted gas oil ratio).

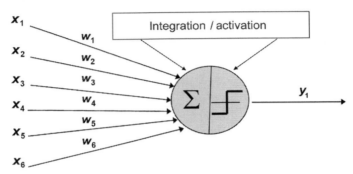

FIGURE 5.7 A typical neuron in an ANN with input (x_i), weights (w_i), and the output (y). (For color version of this figure, the reader is referred to the online version of this chapter.)

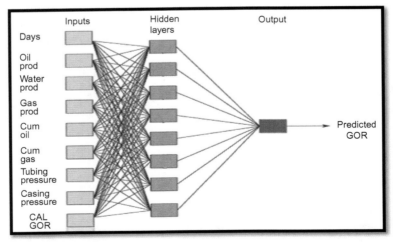

FIGURE 5.8 A simple ANN designed to relate the production data to predicted GOR. (For color version of this figure, the reader is referred to the online version of this chapter.)

5.4.1   Training/Learning

An ANN much like the human brain goes through the training and learning process. The learning possibility of *ANN is one element that has made it an attractive tool*. If we ask an AAN to perform a specific *task* to solve and provide it with *class* of functions *F*, learning means using a set of *observations* to find the right set of functions to establish an *optimal* relationship between the input and the observations. This is done by defining a cost function comprised of the difference between the ideal output and the ANN output and constantly adjusting the weights such that the difference is minimized. For example, in a back-propagation ANN, the weights are recursively updated. Equation (5.5) shows how the ANN weights at time $t+1$ are updated from those at time t using the difference between the desired output d and the ANN output y at time t. Here, ε (between 0 and 1) is the gain or "learning rate" and x_j is the jth input.

$$W_j(t+1) = W_j(t) + \varepsilon(d-y)x_j \tag{5.5}$$

The following are different steps in the training of the ANN:

1. Normalizing, filtering or smoothing data sets if needed.
2. Divide data into three sets:
   i. Training set
   ii. Validation set
   iii. Testing set
3. Train network using the back-propagation algorithm on the training set.
4. The network performance on the validation set is used to determine when to stop training.
5. The testing set is used for prediction and final error measure.

There are a large number of ANN structures such as multilayer perceptron, radial basis function, self-organizing (Kohonen) network and modular neural networks, or Committee Machines. Figure 5.9 shows the configuration of the latter. In this type of ANN, several independent networks, characterized by Expert 1,...,*N*, create their respective outputs. The global expert (gate keeper or Committee Chair) assigns suitable weights to the output of each. It then sends the results to an "integrator" to generate the final output. For more details see Aminzadeh and de Groot (2006).

5.5   FUZZY LOGIC

FL is a computational tool that deals with the linguistic and qualitative nature of information to a computer. It is generalization of the classical or Aristotelian logic of "A thing either is or is not." FL, goes beyond the rigid boundaries of "black" or white and allows the gray area which is the more realistic situation in many cases. The binary language of Boolean algebra, used by nearly all types of modern digital computers, is based directly on the true and false logical variables of conventional logic. Using this logic, computers are able to manipulate precise facts that have been reduced to strings of zeros and ones. The multivalued nature of FL has been employed to allow computers to deal with the "real world" vagueness associated with linguistic and qualitative information.

Through the introduction of the "membership function" concept $\mu(x)$ the degree of belonging of the variable x to a given set allows tremendous flexibility in representing imprecise data and linguistic rules. The characteristics, or shape, of membership functions can be chosen based on mathematical convenience or how accurately they describe a linguistic or physical phenomenon. Figure 5.10 shows three typical membership functions, triangular, trapezoidal, and Gaussian. We also show how these functions can describe physical properties with linguistic qualifiers like porosity of "about" 2%, porosity "approximately between" 4% and 6%, and porosity of "roughly" 13%.

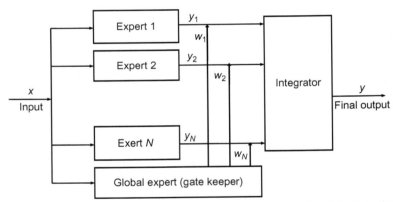

FIGURE 5.9 Configuration of a Committee Machine. *From Aminzadeh and de Groot (2006). Courtesy of EAGE.*

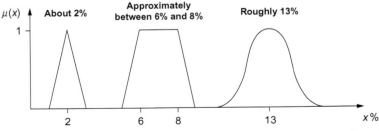

FIGURE 5.10 Membership functions representing different linguistic qualifiers such as about, approximately between, and roughly. *From Aminzadeh and Wilkinson (2004). Courtesy of EAGE.*

5.6   PERMEABILITY PREDICTION FROM LOGS

To illustrate how FL works, we use an example from Aminzadeh and Wilkinson (2004) and Tamhane et al. (2002) for predicting permeability from well logs. Conventional approaches for permeability predictions can be improved if we are able to handle qualitative information based on linguistic descriptions (e.g., "low," "medium," and "high") which are commonly used by experts. We can use supervised fuzzy clustering (FC) techniques to accomplish this. For each unclassified input vector, the membership value of each of the clusters is calculated by a nonlinear transformation of the distance between the vector and the centroid of the cluster. As in NNs, the cluster with the highest membership value is the final classification.

To handle the highly overlapping clusters, we construct a set of hypercluster for each cluster using fuzzy c-means in order to capture the complex distribution of the training patterns. When the hyperclusters of different clusters are overlapped, the algorithm repeats within the overlapped regions until no more overlapped regions are found. Hence, a large number of rules are required to characterize a complex system. The supervised fuzzy cluster boundaries are fuzzy and as such end values of each cluster can be represented as belonging to multiple clusters with different membership values. As such it provides realistic representation of permeability values with respect each cluster.

Four well logs: gamma ray (GR), density (RHOB), neutron (NPHI), and deep resistivity (RT) were available. The permeability quality profile (Fig. 5.11) was constructed by segmenting the WR curve (301 points) into four classes using the three cut-off values of 1, 5, and 10 md. For FC, normalized inputs for the 27 training patterns were used. The hypercluster generation algorithm produced a total of 14 hyperclusters. After clustering, an eight-input ANN was used combine the linguistic indicators and the well logs for predicting permeability values.

Two specific advantages of FL used in combination with NN are in its ability to refine prior predictions and to incorporate experts' expectations. For instance, we may make use of the permeability values from conventional approaches by treating them as prior predictions. By truncating these prior values into linguistic indicators, we may use the proposed NN to refine the accuracy of the predictions. A closer look showed that the mean square error of the predictions (Fig. 5.11) compared to the core permeability data (target data) was 34% smaller than that of the conventional method.

The other advantage of the fuzzy approach is its flexibility to incorporate geologists' expectations. It allows geologists to "edit" the linguistic indicator profile resulted from the clustering techniques based on their geological knowledge and field experience. This is a significant improvement over the previous methods in which the performance is completely driven by the information embedded within the training set. This could be inappropriate if the

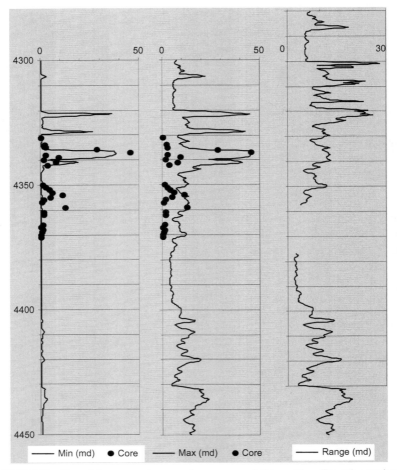

FIGURE 5.11 The minimum, maximum, and range permeability profiles. (For color version of this figure, the reader is referred to the online version of this chapter.)

training set is small, biased, and noisy. The proposed computing framework is general and useful in other real-life engineering problems.

It was concluded that the accuracy of the linguistic descriptions plays an important role for permeability prediction at various depths. The integrated system could refine prior predictions and provide a significant room for incorporating linguistic expectations into the system.

5.7 GENETIC ALGORITHMS

Earlier, we noted that NNs attempt to mimic biological neurons (human brain). GAs that are a class of evolutionary computing (EC) technique are designed to replicate the Darwinian or natural evolution process. EC and its most common element, GA was formally introduced by Holland (1962). The majority of GA applications are focused on performing optimization of search algorithms or machine learning. The latter application has also been extended for parameter learning in ANNs as well as determining membership functions in fuzzy systems. GAs are normally considered as general-purpose stochastic optimization methods for solving search problems. Since many new terminologies will be used to describe foundations of GA, a glossary is included at the end of this chapter with necessary definitions.

GAs start with an initial population (of strings), to search for a number of maximum (or minimum) points (peaks or troughs) in parallel. A genetic operator is capable of exchanging information between many locals, thus lessening the possibility of ending at a local minimum and missing the global minimum. It works with a coding of parameters (strings), not the parameters themselves. Unlike many optimization approaches that require calculation of the derivative of the objective function, GAs only evaluate the objective function (fitness function) for different population sets. The only available feedback from the system is the value of the performance or "fitness" measure.

For example, as described in Caers (2003), we start with $(d_1, d_2, H, h, V_p, V_s)$ as our search space (such as in an elastic inversion problem). The extremes of our search space are given by (120, 200, 3000, 200, 2000, 1500) and (150, 250, 3500, 300, 3000, 2500). An individual point in the search space (111.3, 302.4, 3321.7, 266.4, 1897.5, 2675.0) can be interchangeably represented by a string of real or binary digits shown in Fig. 5.12 (top and middle). The binary codes are derived from replacing the integer values of each element with a string of 5 binary numbers between 00000 and 11111 (between 0 and $2^N - 1$ or 31 for $N = 5$) depending upon the relative value of each of the above elements (e.g., h) with respect to their placement between their respective integer maximum and minimum values. The gray code alternative shown at the bottom of Fig. 5.12 is sometimes used instead of the binary code with the advantage of only one bit being different for the adjacent numbers. This eliminates major problems when crossover and mating takes place (inheriting genes).

5.7.1   Reproduction

GA mimics natural selection process by creating a new population (children or solution points) from the old ones (parents or prior iteration). Through

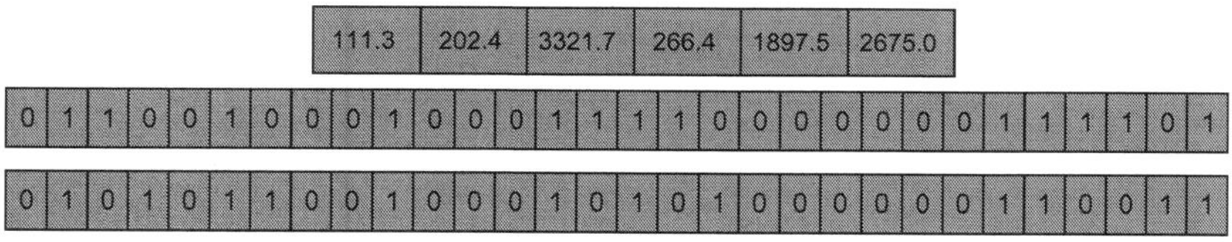

| 111.3 | 202.4 | 3321.7 | 266.4 | 1897.5 | 2675.0 |

| 0 | 1 | 1 | 0 | 0 | 1 | 0 | 0 | 0 | 1 | 0 | 0 | 0 | 1 | 1 | 1 | 1 | 0 | 0 | 0 | 0 | 0 | 0 | 0 | 1 | 1 | 1 | 1 | 1 | 0 | 1 |

| 0 | 1 | 0 | 1 | 0 | 1 | 1 | 0 | 0 | 1 | 0 | 0 | 0 | 1 | 0 | 1 | 0 | 1 | 0 | 0 | 0 | 0 | 0 | 0 | 1 | 1 | 0 | 0 | 1 | 1 |

FIGURE 5.12   Integer, binary, and gray code representation of a point in the solution space. (For color version of this figure, the reader is referred to the online version of this chapter.)

combining the existing strings such as "crossover" that is changing the value of the string at a given point, a new population is produced. For example, in the last line of string in Fig. 5.12, a crossover at the 4th element from the right produces a new string, with the last four elements being 1011. Practically, reproduction is done by combining different genomes according to their pre-assigned fitness values. Among the specific mechanism of the choice of strings to be combined is the "Roulette wheel parent selection." The area of different segments in the roulette (each representing a chromosome or genome) is uneven. The size of those areas is proportional to their respective "fitness," ensuring the "survival of the fittest." Thus, it is more likely for chromosome to survive and contribute to the creation of an offspring with even larger fitness values.

Mutation is another process that leads to generation of new strings. Unlike the crossover that virtually copies pieces of the string from parents, mutation introduces completely new "generation" or fresh blood, thus help further enhancing the population. This is done by switching a randomly chosen bit in the string randomly (with a low probability). Thus, three parameters, population size, n, crossover probability p_c, and mutation probability p_m, combined with the previously defined generation gap, G that allows overlapping population of succeeding generations are the key parameters in a GA problem.

5.7.2　Workflow of a GA

GAs used for many practical problems attempt to mimic the evolution theory described earlier. They all include five key components involving: establishing an initial population, evaluation criteria, reproduction operators, selection criteria, and termination criteria. Figure 5.13 shows a typical implementation of a GA.

Given a way or a method of encoding solutions of a problem into the form of chromosomes and given an evaluation function that returns a measurement that provide the cost value of any chromosome in the context of the problem. The encoding mechanisms and the evaluation function form the links between the GA and the specific problem to be solved. The technique for encoding solutions may vary from problem to problem. Generally, encoding is carried out using bit strings. The coding that has been shown to be optimal is binary coding (Holland, 1975). Intuitively, it is better to have a few possible options for many bits than to have many possible options for a few bits. An evaluation function in a GA plays the same role as the environment plays in natural evolution.

5.8　INTEGRATION OF SC METHODS WITH EACH OTHER AND CONVENTIONAL GEOSTATISTICAL METHODS

Earlier, we discussed CM as well as the unconventional methods. We referred to the latter as SC, comprised of NN, FL, and GAs. Each of these techniques

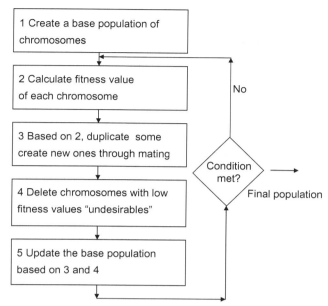

FIGURE 5.13   Basic steps of a genetic algorithm.

has its own strength and weakness. This, however, is very much dependent on the actual problem to be solved and the nature of the input data, information, and the desired output. Based on the surveys conducted, the open literature and perhaps authors personal bias. Table 5.2 provides the "Score Card" for each method for specific real-life problems and challenges. This is done by giving an, albeit subjective, score from A to D in dealing with different types of problems and issues.

Based on the "Score Cards" of Table 5.2, CM or geostatistical-based approaches (CM) are most suitable when a good mathematical model or probability distribution is either in existence (such as physical laws) or can easily be derived or approximated. Also, for many real-time operations such as guiding an airplane or controlling flow of oil in a pipeline that can be modeled through a recursive model updating method (such as Kalman filtering), CM would deliver satisfactory results. In many other situations like when we need linguistic manipulation, handling uncertainty and nonlinearity, ability to learn, handling faulty data, or incorporating or representing human knowledge (especially nonnumerical ones), CM performs poorly.

Conventional statistical methods CM are usually based on sound and well-known mathematical concepts. However, for the most part they are based on very rigid assumptions (e.g., Bayesian probability distribution in many situations or independence of parameters when using regression). They are also not versatile enough to handle nonnumerical data.

TABLE 5.2 Score Card for NN, FL, and GA

Issue	Method			
	CM	*NN*	*FL*	*GA*
Dealing with nonlinearity	D	A	A	A
Expert knowledge	C	B	A	D
Fault tolerance	C	A	A	A
Handling imprecision	D	A	A	A
Knowledge representation	D	C	A	C
Learning ability	D	A	D	B
Linguistic manipulation	D	B	A	D
Mathematical model	A	D	B	D
Optimization capability	C	B	D	A
Real-time operation	A	B	A	C

From Aminzadeh and de Groot (2006). Courtesy of EAGE.

Neural networks, has many advantages, especially in dealing with uncertainty and nonlinearity, fault tolerance and most importantly the ability to learn. However, NN is not very effective in utilizing existing mathematical models or statistical information. It is also not very good for knowledge representation.

Fuzzy logic shines in linguistic manipulation and knowledge representation. It also performs well in handling uncertainty and real-time operation. As an example, there are many FL-based controllers that operate trams, air conditioners, or video cameras. On the other hand, the learning and optimization capability of FL is not very good.

Genetic algorithms main strengths are in optimization of complex objective functions, their fault tolerance, and handling uncertainty. They do not perform well in knowledge representation, linguistic manipulation, and mathematical modeling. GA would perform well to reach a global minima but the convergence to a local minima may not be very fast.

Given the above key differences among intrinsic strength and weakness of different SC methods, integration or combining of one or more of these methods, also referred to as "hybrid" approaches can be attractive. That is, one can combine any two or more SC methods or any SC method with CM method to improve results. Several books have been written on the subject as well. Among those for general applications are Lin and Lee (1996) and Jang et al. (1997) and for the oil and gas applications, Nikravesh et al. (2003) and Aminzadeh and de Groot (2006).

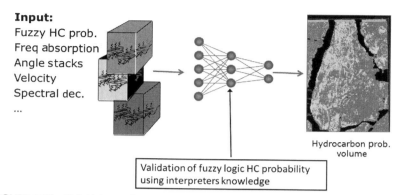

Input:
Fuzzy HC prob.
Freq absorption
Angle stacks
Velocity
Spectral dec.
...

Hydrocarbon prob.
volume

Validation of fuzzy logic HC probability
using interpreters knowledge

FIGURE 5.14   Hybrid (neuro-fuzzy) method to estimate the hydrocarbon probability volume. (For color version of this figure, the reader is referred to the online version of this chapter.)

Combining different SC techniques with each other and/or with the conventional statistical techniques to characterize reservoirs or to predict presence of reservoirs has been done extensively. As an example, Fig. 5.14 from Aminzadeh and Brouwer (2006) shows the result of a case history where a neuro-fuzzy approach was used to high-grade prospects with the highest probability of hydrocarbon (HC) occurrence.

The procedure involves creating a preliminary HC probability volume based on different (fuzzy) rules of thumb on the ranges of different seismic attributes that may be related to the presence of HC. The fuzzy HC probability volume, along with a number of seismic attributes (as discussed in Chapter 3) is provided as an input to a NN. Some of the seismic attributes derived from a 3D seismic volume are frequency absorption, partial (angle) stack gathers, seismic velocity volume, and spectral decomposition volume (see Chapter 3 for the definition of all these attributes).

The NN is trained by some input from the interpreter on the potential areas of HC. In addition, if any exploration well or another nearby well information is available, could be used as input to the NN. After several iterations, a HC probability volume is generated. A horizontal slice of the volume is shown in Fig. 6.14, where the lighter color (yellow) indicate a higher probability of HC occurrence.

REFERENCES

Aminzadeh, F., Brouwer, F., 2006. Integrating neural networks and fuzzy logic for improved reservoir property prediction and prospect ranking. In: 76th Ann. Internat. Mtg: Soc. of Expl. Geophysicists.

Aminzadeh, F., Chatterjee, S.L., 1984. Application of clustering in exploration seismology. Geoexploration 23, 147–159.

Aminzadeh, F., de Groot, P., 2006. Neural Networks and Other Soft Computing Techniques with Applications in the Oil Industry. EAGE Publications, The Netherlands, pp. 13 and 16–23.

Aminzadeh, F., Wilkinson, D., 2004. Soft Computing for qualitative and quantitative seismic object and reservoir property prediction, Part 2, Fuzzy logic applications. First Break, EAGE 22, 69–78.

Caers, J., 2003. Geostatistics: from pattern recognition to pattern reproduction. In: Nikravesh, M., Aminzadeh, F., Zadeh, L.A. (Eds.), Soft Computing and Intelligent Data Analysis. Developments in Petroleum Science, vol. 51. Elsevier, Amsterdam, pp. 97–115.

Dasgupta, S., Kim, J., Al Mousa, A., Al Mustafa, H., Aminzadeh, F., Lunen, E., 2000. From seismic character and seismic attributes to reservoir properties: case study in Arab-D reservoir of Saudi Arabia. In: 70th Ann. Internat. Mtg: Soc. of Expl. Geophys., pp. 597–599.

Deutsch, C.W., Journel, A.G., 1998. GSLIB Geostatistical Software Library and User's Guide. Oxford University Press, Oxford, UK.

Holland, J.H., 1962. Genetic Algorithms. Scientific American, Jul'92, pp. 66–72.

Holland, J.H., 1975. Adaptation in Natural and Artificial Systems. University of Michigan Press, Ann Arbor, MI.

Jang, J.-S.R., Sun, C.-T., Mizutani, E., 1997. Neuro-Fuzzy and Soft Computing: A Computational Approach to Learning and Machine Intelligence. Prentice Hall, New Jersey, USA.

Kaiser, M.J., Yu, Y., 2012. Oil and gas company valuation, reserves and production. Oil Gas Financ. J. 9 (3).

Lin, C.T., Lee, C.S.G., 1996. Neural Fuzzy Systems. Prentice Hall, New Jersey, USA, 797pp.

Mohaghegh, S.D., 2000. Virtual intelligence and its application in petroleum engineering: Part 1. Artificial neural networks. J. Pet. Technol. 52, 58046, Part 2. Evolutionary computing, Part 3. Fuzzy logic: JPT, SPE Papers, 61925, 62415.

Nikravesh, M., Aminzadeh, F., Zadeh, L.A., 2003. In: Soft Computing and Intelligent Data Analysis. Developments in Petroleum Science Series, vol. 51. Elsevier, UK, 724pp.

Tamhane, D., Wong, P.M., Aminzadeh, F., 2002. Integrating linguistic descriptions and digital signals in petroleum reservoirs. Int. J. Fuzzy Syst. 4 (1), 585–591.

Wong, P.M., Aminzadeh, F., Nikravesh, M., 2002. Soft Computing for Reservoir Characterisation and Modeling. In: Studies in Fuzziness and Soft Computing, Physica-Verlag, Springer-Verlag, pp. 3–12.

Reservoir Characterization

INTRODUCTION

Accurate reservoir characterization is a key step in developing, monitoring, and managing a reservoir and optimizing production. To achieve accuracy and to ensure that all the information available at any given time and is incorporated in

Developments in Petroleum Science, Vol. 60. http://dx.doi.org/10.1016/B978-0-444-50662-7.00006-8
151

the reservoir model, reservoir characterization must be dynamic. To achieve this goal, however, one starts with a simple model of the reservoir at a given time point (a static model). As new petrophysical, seismic, and production data become available, the reservoir model is updated to account for the changes in the reservoir. The updated model would be a better representative of the current status of the reservoir. Both static reservoir properties, such as porosity, permeability, and facies type; and dynamic reservoir properties, such as pressure, fluid saturation, and temperature, needs to be updated as more field data become available. Characterizing a reservoir by updating of both static and dynamic reservoir properties during the life of the field is referred to as dynamic reservoir characterization. Dynamic reservoir characterization is discussed in Chapter 7, dealing with time lapse or 4D geophysical data and reservoir monitoring. This chapter, however, focuses on static reservoir characterization.

In this chapter, we will focus on two aspects of reservoir characterization, namely, static reservoir characterization and geophysics for reserves and resources estimation. The latter has become more important recently with the new regulations and guidelines on reserves and resource certification where geophysics is expected to play an important role in improving such estimates and reducing the associated uncertainties.

The main objective of reservoir characterization is to transform the available seismic, log, geological, production, and other data to reservoir properties. The reservoir properties include: reservoir thickness, number of reservoir units, porosity, permeability, pressure distribution, fracture distribution (in the case of unconventional reservoirs), and fluid saturation (oil, gas, water). Different people may have a different notion of "reservoir characterization." The SLB oil field glossary (http://www.glossary.oilfield.slb.com) gives the following definition:

A _model_ of a _reservoir_ that incorporates all the characteristics of the reservoir that are pertinent to its ability to store hydrocarbons and also to produce them. Reservoir characterization models are used to simulate the behavior of the fluids within the reservoir under different sets of circumstances and to find the optimal _production_ techniques that will maximize the production.

This is followed by yet another definition:

The act of building a _reservoir_ model based on its characteristics with respect to fluid flow.

Reservoir characterization usually refers to a static reservoir model. However, for many applications, including reservoir monitoring, having a dynamic reservoir model is more desirable.

6.1 OVERVIEW

Most reservoirs are much more complex than as originally considered. A common saying in reservoir description is "reservoirs can only be understood backwards but they are produced forwards." Geologists, geophysicists, and engineers usually have a different view of "reservoir characterization." Geologists view reservoir characterization as extrapolation of different geological data such as outcrops, core samples, and geochemical data geologic maps), to assess and predict deeper reservoir properties. Schatzinger and Jordan (1999) at AAPG Memoir 71, include many original ideas on different aspects of reservoir characterization.

Schatzinger and Jordan (1999) begin by the following: reservoir characterization is the process of creating an interdisciplinary high-resolution geosciences model that incorporates, integrates, and reconciles various types of geological and engineering information from pore to basin scales. They consider maintaining high displacement efficiency, optimizing high sweep efficiency, providing reliable reservoir performance predictions as well as reducing risk and maximizing profits as the main goals of reservoir characterization. Figure 6.1 depicts one such view of reservoir characterization.

The engineer's view of reservoir characterization is to extrapolate well data and production data as well as reservoir simulators, to extrapolate reservoir properties away from the existing wells. For example, sometimes, only production data are used for characterizing reservoirs (e.g., Gaskari and Mohaghegh, 2007).

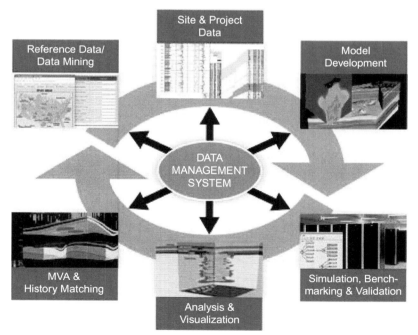

FIGURE 6.1 A view of "reservoir characterization" for carbon sequestration application, from DOE's Pacific Northwest National Laboratory: http://csi.pnnl.gov/articles/g/e/o/Geologic_Sequestration_Software_Suite_80fc.html. (For color version of this figure, the reader is referred to the online version of this chapter.)

The detailed spatial coverage offered by geophysical data is calibrated with analysis of well logs, pressure tests, cores, geologic depositional knowledge, and other information from appraisal wells. Geophysicists primarily use seismic data to perform seismic inversion and relate acoustic and elastic impedance as well as tuning thickness, frequency attributes, amplitude versus offset (AVO) and other seismic indicators to derive porosity, bed thickness and possibly saturation, reservoir pressure, and permeability. One goal is to use seismic data to help with the "extrapolation" process to the entire field from well data.

Whether we take the geologists, engineers, or geophysicists view of reservoir characterization, ultimately, the goal is to integrate all the available data to deduce different properties of the reservoir. Preferably, the integration should be done in a "proper" manner. That is to ensure the integration is done in the discipline level rather than integrating the results derived from different experts having worked in isolated discipline-based organizations (Fig. 6.2).

It must be emphasized that often characterization of a reservoir is an inexact process that begins with interpretation of information, followed by a description of the heterogeneity of reservoirs and an approximate formulation of mathematical models of the complex reservoir behavior. The advent of 3D seismic data has been a major factor in advances in reservoir characterization. A correct initial geological description followed by building reservoir model and finishing numerical simulation models would be the natural sequence of events for reservoir characterization.

6.2 SURE CHALLENGE

We established earlier that reservoir characterization is based on the integration of different data types. A reservoir's life begins with exploration that leads to discovery, which is followed by delineation of the reservoir,

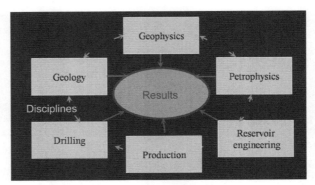

FIGURE 6.2 Proper integration (discipline vs. results). (For color version of this figure, the reader is referred to the online version of this chapter.)

development of the field, production, and finally abandonment. Integrated reservoir management is imperative to a successful operation throughout the life of the reservoir. The entire process of exploration for reservoirs to its abandonment involves acquisition and analysis of different types of data. These data are associated with an enormous range of scale as shown in Fig. 6.3. This spans ultrasonic measurements of pores of the order of 1 mm to remote sensing measurements of basins of over 10 km wide. Examples of many other data measurements for many other objects that lie in between the above-mentioned features are shown in this figure.

Admittedly, not all the data types are integrated at the same time. Nevertheless, the wide range of scale differences for different data types is one of the challenges in reservoir characterizations. To make the matter more complicated is the fact that different data types are associated with different levels of uncertainties. For example, the direct measurements of rock properties from the core data may involve little uncertainty. The petrophysical information from well log data may be associated with somewhat more uncertainty. The seismic data used to ascertain reservoir properties, for their indirect nature of measurements, involve much more uncertainty. Thus, uncertainty level and its variations with respect to different data types are another challenge.

Resolution is yet another challenge. The resolving power of different data types is drastically different. As shown in Fig. 6.3, some data types have very high resolving power. For example, while well log data can resolve a reservoir

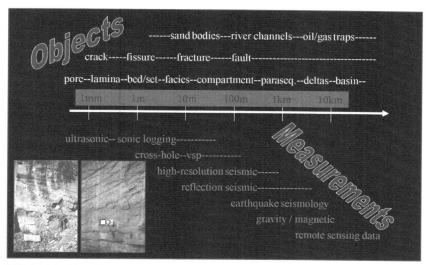

FIGURE 6.3 For different data types associated with different geological and reservoir features wide range of physical scale. (For color version of this figure, the reader is referred to the online version of this chapter.)

unit of under a foot, seismic data may not be able to resolve a reservoir under 30 ft. Finally, the effectiveness and usefulness of different data types are impacted by the geological and reservoir "environment." This can be associated with different types of reservoir types (carbonate, clastic, unconventional, high pressure/high temperature, ultra deep water, heavy oil, etc.).

We refer to these four key issues: scale, uncertainty, resolution, and environment as: SURE challenges. Figure 6.4 illustrates three key data types: core, well log, and seismic data can be considered as data pyramid (on the left hand side). The base of the pyramid is the seismic with very large coverage but with limited resolution and lesser level of certainty. The top of the pyramid is the core data with very little coverage (only at a particular well location involving a fraction of the well) but with high level of certainty and resolution. Effective integration of all the data types, in spite of the SURE challenge is what reservoir characterization is all about.

The right hand side pyramid in Fig. 6.4 (the upside down one) shows a different aspect of integration. That is, vast amount of data needs to be combined with some technical knowledge and experience to perform effective data mining and ultimately reservoir characterization. As an aside, it must be pointed out that borehole geophysical data (e.g., vertical seismic profile and crosswell data) can locally help fill the gap between well log and 3D seismic. In addition, microearthquake (MEQ), or passive seismic data, in conjunction with the conventional seismic data are also used to do reservoir characterization. We will provide more details on microseismic data with a special focus on their applications in unconventional reservoirs and fracture characterization in Chapter 9.

Uncertainty in hydrocarbon volumes is typically assessed through multiple realizations of the reservoir model. Many models can be generated using the same set of interpreted data. This is mitigated by integration of disciplines that limits the possible model realizations and ultimately makes the model unique.

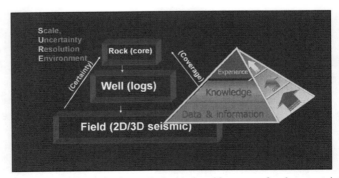

FIGURE 6.4   SURE challenge: having to deal with the wide ranges of scale, uncertainty, resolution, and environment of different data types when integrating them. (For color version of this figure, the reader is referred to the online version of this chapter.)

6.3   INTEGRATION OF THE DISCIPLINES

Earlier, we talked about the challenges of integrating different data sets, highlighted by what we refer to it as the SURE challenges. Since data integration is the key element of reservoir characterization, we will elaborate further on this topic. As illustrated in Fig. 6.5, integration of the disciplines is the key to managing the risk. The objective is to integrate all the available subsurface data and deduce different rock and fluid properties of the reservoir. Rather than simply integrating the interdisciplinary data and results derived from experts working in isolation, integration at the discipline level is a more robust approach. Forecasting the recovery potential and production performance of a field requires understanding of the geologic framework and the petroleum system. This is achieved through the integration approach.

Estimates of oil and gas volume in the reservoir are obtained from analysis of geological and geophysical data. Reservoir engineers use this estimate for field development and for production planning in the economic and efficient recovery of the hydrocarbons. The process starts with reservoir discovery from exploration drilling. This is followed by appraisal and development drilling that leads to the field development plans and for making production decisions. The formulation of development plan for the reservoir defines the number of wells and locations of production and injection wells to be drilled. The imaging necessary to optimally develop hydrocarbon reservoirs far exceeds the details and accuracy required in exploration for discovering them. This precept has resulted in expanding application of geophysical techniques

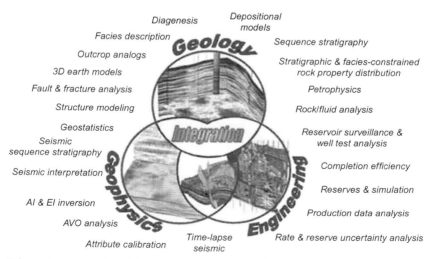

FIGURE 6.5   Integration of the disciplines—experts working together with different data types ensure more robust approach to delineation and management of reservoirs. *Courtesy: iReservoir Inc. (www.ireservoir.com/images).* (For color version of this figure, the reader is referred to the online version of this chapter.)

especially 3D seismic analyses with different degrees of importance from the exploration phase all the way to reservoir management.

Engineers forecast a production profile (production rate vs. time) also referred to as the decline curve, for the reservoir depletion and an injection profile if water or gas is to be injected. At this stage, it is necessary to determine the reservoir drive mechanism and the size and strength of the aquifer. Hydrocarbons are produced through displacing them from reservoir pore spaces by aquifer water or gas. The gas drive may be from expansion of the gas cap as the reservoir pressure declines, or dissolved gas in oil as it comes out of solution with pressure decline.

6.4 GEOPHYSICS IN RESERVOIR CHARACTERIZATION

Geophysics contributes to reservoir characterization and its management (monitoring) by adding value (improvement in production plan) and by minimizing risk—drilling risk of dry hole, risk of blow out, risk of inefficient recovery process, among others. Thus, geophysics can be considered as a risk reduction tool, it reduces exposure to loss. Geophysical analyses can save costs by reducing the drilling risk and/or reducing dry holes and poor producers. Geophysics can contribute to reservoir economics by adding reserves and by reducing drilling cost. Robertson (1989) suggested that geophysics impacts reservoir management in two ways:

1. Geophysical reservoir study adds hydrocarbon reserves that would not be produced by the existing development plan. Bypassed pay in untapped reservoir compartments could be identified from interpretation of three-dimensional (3D) seismic data.
2. Geophysical analysis could provide data for improved reservoir surveillance for fluid-flow monitoring. Geophysical techniques could add quantitative information for enhancing and constraining reservoir simulation models.

Applications of geophysical data for reservoir monitoring and management will be further discussed in Chapter 9.

Geophysical data provide measurements between the wells to extend geologic description away from the wellbore and therefore plays an important role in developing a reservoir model. As described earlier, reservoir architecture (structure) and the reservoir properties are derived from the analysis and integration of data from various geoscience disciplines. The distributions of the reservoir and non-reservoir rock types and of the reservoir fluids determine the geometry of the model and influence the type of model to be used. For example, the number and scale of the shale breaks (or dense carbonate) in reservoir geometry determine the continuity of the reservoir facies and influence the vertical and horizontal dimensions of each model cell. Incorporation of geologic model into a reservoir model requires recognition of detailed reservoir heterogeneity.

Multidomain data from geophysical, geologic, petrophysical, well test, and production history data are integrated in developing reservoir models. The geophysical data integrated with well logs and reservoir fluid data provide: (1) static reservoir model and (2) continuously evolving dynamic monitoring. The static reservoir model provides a representation of the structure, thickness, lithology, porosity, initial fluids in the reservoir. Dynamic characterization is a representation of the fluid flow in a static reservoir model and needs to be updated and validated periodically with reservoir performance data—pressure changes, production, and injection rates.

Geophysical measurements and geological observations are integrated to optimize reservoir development and production. To accomplish this, data are constantly being evaluated to form the basis for locating production and injection wells, managing pressure maintenance, and for performing workovers. These activities generate new data—logs, cores, DSTs that change the maps, revise the structure, and alter the reservoir stratigraphic model. Figure 6.6 shows a workflow using geophysical data analyses and their interpretation by calibration with measurements in wells. The process provides an integrated subsurface model that incorporates spatial data between wells. Reservoir structure and rock properties are derived from this process.

Integration of 3D seismic data with well logs and core analysis data allows accurate estimation of reservoir volumetrics, fluid properties, and lithology.

FIGURE 6.6 Workflow showing analysis and integration of geophysical data for subsurface modeling. *Courtesy of iReservoir Inc (www.ireservoir.com).* (For color version of this figure, the reader is referred to the online version of this chapter.)

Geophysical techniques are also being applied to monitor and control of producing wells. Geophysical measurements are used in monitoring of CO_2 injection, gas injection, gas contact movement, hydraulic fracturing in unconventional reservoirs shale gas and oil, thermal systems such as steam injection, and *in situ* combustion for heavy oil production, oil–water systems including primary depletion, natural water drive, and water injection. In addition, seismic emissions associated with stress changes in and around the reservoir can be used to image the reservoir dynamics.

6.4.1 Reservoir Characterization: The Key to Reservoir Management

A detailed description of the reservoir rocks, fluids, and the aquifer is essential to optimize hydrocarbon recovery. The need for reservoir description begins as soon as the discovery is made in order to estimate the hydrocarbon in place, recoverable reserves, and rates of production. Usually, as a field or reservoir goes through the life cycle of appraisal, planning, development, and surveillance, an ever more complete reservoir description is necessary. Optimum reservoir management requires teamwork and close coordination among geologists, geophysicists, engineers, and managers through all stages of the life of a reservoir. Geological and geophysical data are essential elements of most aspects of reservoir description. This provides us with information on the reservoir facies, the qualitative and quantitative reservoir rock properties, reservoir rock fabric—lithology, porosity, and permeability distribution, expected to be encountered. The reservoir engineers use this data in planning the development well locations so that they connect to the best porosity development.

Determination of reservoir continuity is an example of reservoir characterization that is necessary. A first step in reservoir description process is the recognition of correlative reservoir subzones or layers, the high porosity compartments of the reservoir and also intervening impermeable, or low permeability strata. Understanding of the depositional environment and the diagenetic processes controlling the reservoir and non-reservoir rock is essential in making these correlations. Figure 6.7 shows a typical outcome a reservoir characterization process where the well log data and seismic data are integrated to display different reservoir properties at different strata or 3D volume of that property. Well tracks with color-coded well log values are overplayed on the porosity volume derived from integrating well logs and seismic data.

Reservoir development must follow a geological model that is refined as development proceeds. Engineering data are as critical as geological data in defining heterogeneity and all available information must be integrated across disciplines to maximize recovery from the reservoir. Production volumes that fail to meet or exceed prediction, pressure anomalies, well-test results that do not conform to predictions are major indications that the geological model is

FIGURE 6.7   A typical reservoir characterization results. *Source: GOCAD software.* (For color version of this figure, the reader is referred to the online version of this chapter.)

not optimum. The reservoir model must be designed in terms of facies building blocks, analyzed in terms of flow continuity, and executed as a development plan for the reservoir that is progressively refined as each well is drilled.

Figure 6.8 shows a 3D seismic volume of a reservoir above a salt dome. The facies map shows a buried meandering channel sand, the background depicts other facies. Without the 3D seismic, such definition and distribution of the subsurface channels would have been impossible, unless a close grid of wells was drilled in the area. Seismic attributes also play a role in relating different facies to different reservoir units.

The reservoir properties in Figs. 6.5 and 6.6 (in these cases porosities and facies) are derived from 3D data seismic data on the basis of the seismic character, away from the well locations with proper extrapolation. The geologic model is divided up into a mesh of grids of 3D blocks which represent the numerical description reservoir. In general, the following steps would be taken to build the reservoir model:

1. 3D seismic horizon interpretation
2. 3D seismic attribute volume interpretation
3. Building 3D reservoir structural model
4. Building 3D common share earth model
5. 3D seismic and log attribute up scaling and gridding
6. 3D integrated reservoir property modeling
7. Integrated reservoir volume property estimation.

Figure 6.9a and b illustrates the above steps.

Reservoir engineers seek to obtain the highest possible economic hydrocarbon recovery from petroleum reservoirs. The objective is to position wells

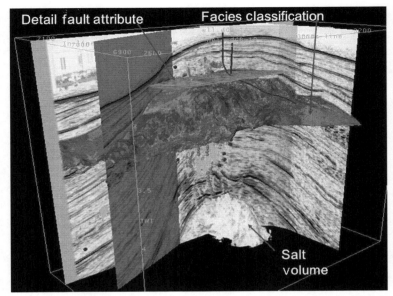

FIGURE 6.8 Visualization of the facies in conjunction with the fault attributes, well track, and the 3D seismic data. *Source: http://OpendTect.org software.* (For color version of this figure, the reader is referred to the online version of this chapter.)

for maximum recovery and later to maximize the volume of hydrocarbon contacted by injected fluids to achieve optimum sweep efficiency. The primary goals in developing a production strategy for a petroleum reservoir are:

1. Optimize the total reserves
2. Reduce the cost of field development
3. Optimize production recovery
4. Minimize operating costs of the developed field
5. Enhance recovery if economically justified.

Two significant challenges that the reservoir engineer faces in order to meet these goals are:

1. Early and accurate characterization of the reservoir properties in terms of volumetric, lithology, rock properties, fluid property, porosity, permeability, reservoir pressure, and its continuity.
2. Improvement in the reservoir surveillance techniques so that the field under production can be efficiently managed. More on the item 2 will be discussed in Chapter 7 on reservoir monitoring.

In order to meet these challenges, more complete and up-to-date information for the reservoir is necessary. The conventional reservoir engineering data— well logs, core analysis—and production history data will have to be

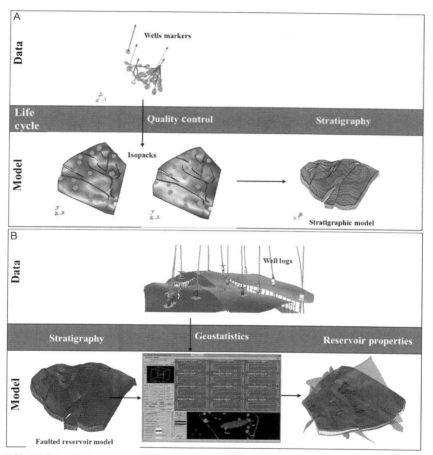

FIGURE 6.9 (a) Demonstration of reservoir characterization process starting with the well marker and going through building reservoir models from ispocks to stratigraphic modeling (GOCAD). (b) Adding more details to the model including faults and using geostatistics (GOCAD). (For color version of this figure, the reader is referred to the online version of this chapter.)

augmented with data away from the wells. This is accomplished through use of the 3D seismic data.

6.4.2   Scales of Reservoir Description

There are four categories of reservoir scales. The largest scale, gigascopic encompasses the entire reservoir. The reservoirs are explored for, discovered, and delineated by incorporating 2D and 3D seismic data and well information at this scale. Models from the smaller scales are "scaled up" to this scale. The megascopic scale is the scale of reservoir simulation. Reservoirs are engineered and managed at this scale. Here, there is a concern for the production behavior around single wells, between well pairs, within well patterns, and over the full field. Along with 3D seismic and wireline logs, inter-well tracer tests, and pressure transient analysis data are used at this scale. Verification of the data is done at this scale. The macroscopic scale is that of log and core analyses. Rock and fluid properties are determined for input into reservoir models and to calibrate well logs and well tests. Finally, there is the microscopic scale which is that of thin section where detailed pore-scale analyses are conducted. Data from these various measurements with different resolution and accuracy are integrated and assimilated in the three-dimensional reservoir models.

During the life cycle of a reservoir, the information needs and the scale of reservoir description changes. As the reservoir matures and the hydrocarbon remaining in place becomes more difficult to recover, fluid movement in the reservoir needs to be closely monitored. Such monitoring will not be possible without improved reservoir characterization. The following Table 6.1 shows the various stages of development cycle of a reservoir and the corresponding geological model, the scale of characterization needed.

Whole reservoir models are developed to study the relative contribution of aquifer influx and support, gas cap expansion, pressure decline, and solution–gas expulsion to the overall energy and mass balance of the reservoir. Between-well models are developed to study how vertical and lateral variations of reservoir flow properties affect the displacement efficiency and volumetric sweep of injected fluids. Details of the models can be increased with improved definition of reservoir characteristics and flow properties. 3D seismic data provide large volumetric coverage with a resolution of 20–200 ft. Sonic logs provide limited volumetric coverage in the immediate vicinity of the well but at much higher vertical resolution 0.3–0.5 ft. Ideally, it would be desirable to have approximately 10-ft resolution throughout the inter-well region, in order to understand the fluid-flow behavior in the region. Crosswell seismic data can be used to fill in the resolution gap locally. However, it is a 2D rather than a 3D technique.

Table 6.1 summarizes the subsurface information required by reservoir engineers and the type of data which can provide the information, for a comprehensive evaluation of the reservoir.

6.4.3 Role of Seismic Data in Reservoir Characterization

Seismic data are used by reservoir management teams to plan and monitor the development and production of a field. Seismic data have the potential to provide the bridge between well logs and core analysis on the one hand, and tracer and well-test analysis on the other. Most geologic maps and models based on 3D seismic data, until now, have been constructed for hydrocarbon exploration rather than recovery. As a result, more emphasis is placed on the reservoir geometry or external configuration and not on the internal fabric that controls the recovery behavior. For reservoir exploitation, models need to explain

TABLE 6.1 Reservoir Description: Subsurface Information Needed

Data Type	Subsurface Information	Discipline	Comments
Seismic	Structural style External geometry Internal fabric	Geophysicist *Engineer* *Geologist*	Seismic data calibrated with well data
Well data	Reservoir Framework Rock properties	Geologist *Petrophysicist* *Geophysicist* *Engineer*	Logs, core data correlated between wells using depositional model
Wireline logs	Net pay	Engineer *Geologist* *Petrophysicist*	Logs correlated by core data
Wireline logs	Fluid contacts	Engineer *Petrophysicist* *Geologist* *Geophysicist*	Logs, well testing and seismic data
Wireline logs	Porosity	Engineer *Petrophysicist* *Geologist*	Logs calibrated by core porosities
Wireline logs	Fluid distribution Water saturation	Engineer *Petrophysicist* *Geologist*	Logs calibrated by oil-based cores or by capillary pressure tests
Cores	Permeability	Engineer *Geologist*	Average core consistent with well test and depositional model
Seismic	Aquifer	Geologist *Geophysicist* *Engineer*	Reservoir data extended to aquifer

reservoir heterogeneity—vertical zonation, lateral compartmentalization and what contributes to the anisotropy or directionality of fluid flow in the reservoir.

The analysis of 3D seismic data can provide information on the geometry of the reservoir, calibrate the rock properties, and flow surveillance. From this information, the recovery strategy is formulated. We make the subsurface maps and models from smooth interpolation of petrophysical data measured at well locations. Well information, however, is unevenly and sparsely distributed. In heterogeneous formations and structurally complex reservoirs, these interpolated models can lead to gross errors with costly consequences on field development. The volume of reservoir rock investigated over a field

TABLE 6.2 Seismic Parameters and Their Seismic Stratigraphic Interpretation

Seismic Parameter	Geological Interpretation
Reflection configuration	Bedding patterns Depositional processes Erosion and paleotography Fluid contacts
Reflection continuity	Bedding continuity Depositional processes
Reflection amplitude	Velocity–density contrast Bed spacing Fluid content
Reflection frequency	Bed thickness Fluid content
Interval velocity	Estimate of lithology Estimate of porosity Fluid content
External form and areal association of seismic facies units	Gross depositional environment Sediment source Geologic setting

by core analysis and wireline logging is on the order of less than one part in a billion. With such a sparse sampling, we need to construct detailed three-dimensional models of the reservoir in order to evaluate and estimate reserves, design an economical and effective drilling program, and optimize resource recovery. Engineers require high resolution, numerical information about the spatial variation, vertically and laterally of a reservoir's fluid-flow properties.

Table 6.2 summarizes the information obtained from the analysis of seismic reflection data and the corresponding geologic interpretation from the analysis.

6.4.4   Geomechanical Properties Characterization

3D seismic data provide information on characterization of geomechanical rock properties at the target formations. For unconventional oil and gas production from shale reservoirs, the geomechanical properties of the rocks are imperative for drilling and well completion by hydraulic fracturing. The seismic attributes are calibrated with measurements of mechanical properties at wells and are used to extrapolate information at well bores to the inter-well regions. Information from well logs and cores and gives a detailed view of the reservoir at well locations. The interpretation provides understanding of rock properties and stress characteristics at the target subsurface. The *in situ* stress state of these rocks can be estimated from seismic elastic attributes.

Knowledge of the stress state prior to drilling is useful for predicting areas at risk for wellbore failure. These properties, therefore, have direct bearing on the placement of wells, reservoir productivity, and the safety issues in fracturing strategy. It is assumed that the rocks *in situ* in the subsurface are constrained horizontally, that is, the horizontal strain is zero in their natural state, and that the rocks are undergoing elastic deformation. Hooke's law describes the relationship between strain and stress. Stress and strain are functions of the elastic properties of rocks. The stress induced during hydraulic fracturing causes sufficient strain on the formation leading to rock failure. Goodway et al. (1997) introduced AVO inversion techniques to derive Lamé's parameters (λ: Lambda, μ: Mu) and density (ρ: Rho) from prestack seismic data. These elastic moduli can be transformed to estimate Young's modulus, Poisson's ratio, bulk modulus, and shear modulus. These moduli are important in estimating how rocks will fracture and whether the fractures will remain open.

In the planning of an optimal hydraulic fracturing program, geomechanical factors such as brittleness and closure pressure are important. These can be estimated between existing wells based on Young's modulus and Poisson's ratio derived from inversion of seismic data. Young's modulus is directly proportional to the brittleness of the rock and the stimulated fracture length. Higher brittleness is an indicator being able to fracture the rock. The *in situ* fracture density can be estimated from various reflection attributes such as amplitude versus azimuth (AVZ) or impedance. The stimulated rock volume (SRV) can be predicted from these properties. AVZ can also provide the magnitude and orientation of local stress field variations and pore pressure variations.

Indicators for optimal hydraulic fracturing often vary over short distances. Rock stress behavior often changes within the space of a few hundred meters, requiring adaptation of the fracturing program. When coupled with more traditional seismic attributes such as acoustic impedance, properly designed, and processed 3D seismic data sets provide many attributes that can be used to model and delineate reservoir heterogeneities that have influence on well performance. Lithological and geomechanical models derived from attributes computed from 3D seismic data can be correlated to predict flow in stimulated wells.

6.4.5   Seismic Data and Geostatistics

The application of seismic reflection technology has traditionally been limited historically to oil field exploration for locating possible traps for hydrocarbon deposits. However, in recent years, geophysical data are playing a key role in reservoir development, production, and EOR projects. As new conventional exploration opportunities become less abundant, the attention, for applications of geophysics shifts to the understanding of subsurface characteristics of unconventional reservoirs and improving the recovery factor of all reservoirs through better monitoring of the process.

The recent years have seen major advances in seismic technology. Improved data acquisition by digital telemetry, better processing techniques by supercomputers, and enhanced interpretation/modeling using interactive graphics workstations are some of the major advances. For the reservoir engineer, the most significant geophysical advance has been the emergence of 3D seismic as a cost-effective tool for describing a reservoir's external and internal structure. More accurate description of reservoir architecture, internal stratigraphy, flow paths, continuity, fluids, and fluid-flow parameters can be obtained.

In contrast to the well data, seismic measurements are relatively imprecise, and their interpretation is somewhat ambiguous. Seismic data, however, provide a spatial density of information that can be incorporated with all other available engineering and geologic data in predicting the reservoir properties away from well control points. A complete description of a reservoir from a petroleum engineering perspective requires measurements over different scales shown in Fig. 6.4.

In the geostatistical approach, the reservoir models are generated by analyzing spatial patterns of correlation between the well and seismic data. As was discussed in Chapter 5, geostatistical reservoir characterization recognizes that spatial correlations may exist in rock properties. The aerially dense seismic attribute measurements are integrated with the sparse well data by "cokriging" a geostatistical technique. This performs a spatial autocorrelation in variation of well log-derived properties and a spatial cross-correlation between those properties and seismic parameters like travel time and amplitudes. By calibrating reservoir properties with 3D seismic inverse modeling results and applying multidimensional geostatistical techniques superior 3D reservoir characterization models can be formulated. Reliability of the models is assessed by comparing the predicted reservoir at a few wells deliberately omitted from the study population. Two case histories at the end of this chapter illustrate some of the steps used to deploy geostatistical concepts in reservoir characterization.

The inherent assumption made in incorporating seismic inversion and calibrating with borehole data is that the relative true amplitudes in the seismic data are preserved and the amplitudes are proportional to the reflection coefficients. These reflection coefficients can be computed from normalized difference of acoustic impedance across the reflection boundaries. Porosity–acoustic impedance crossplot over the reservoir window is used to define and empirical relationship between porosity and impedance. This is then used to invert the seismic-derived acoustic impedance. This information and other well parameters affecting the acoustic impedance data have to be interpolated between borehole locations. If these parameters are highly spatially variable, then any error in their lateral prediction would reduce the accuracy of the seismically derived variable. The porosity derived represents a gross average over the reservoir interval (some compensation is required to infer net porosity thickness).

A significant improvement in reservoir characterization is achieved by integrating 3D seismic data with well log, core, and geologic information. The seismic measurements provide a more detailed 3D image of the geometry, stratigraphy, and faulting than would be possible with well data or seismic data, alone.

Engineers require high resolution, numerical information about the spatial variation of a reservoir's fluid-flow properties. Well logs and core analyses provide high resolution, microscopic, and macroscopic data close to the borehole. The problem is how to interpolate these data between well control and relate the wells to inter-well reservoir properties and scale up the model to the field size or gigascopic models?

6.4.6   Introduction and Evolution of Seismic Attribute

As discussed in Chapter 3, seismic attributes have been used implicitly or explicitly since the introduction of the seismic signal processing by Treitel and Robinson in the early 1960s. The fundamental reason for any type of signal processing/analysis is to understand the nature of the signal and derive some information from it regarding the underlying principles and phenomenon that it relates to. This applies to the signal associated with the engine in a car inspection, the electrocardiogram signal of a heartbeat, passive seismic measurement related to an earthquake, or the seismic measurements made from an active or "controlled" seismic source which is the main topic of this study.

Experts in each field can visually inspect such signal and make an assessment of the situation whether it is a problem with car engine, a particular anomalous heart beat, major change in resistivity, indicating a possible pay zone, occurrence of the earthquake or a major impedance contrast represented by a strong amplitude of the seismic response. Going beyond the visual inspection of the data, for either a more thorough and in-depth analysis of the signal or for more automated analysis of the data, one may develop alternative measures of the signals for specific applications.

Seismic attributes (as discussed in Chapter 3) are certain transformation of the original seismic data to highlight specific features of the reservoirs. The simple acoustic impedance introduced in Chapters 3 and 4 can be considered as a simple seismic attribute. In the early (late 1960s–early 1970s) days of the seismic data-based exploration, "bright spots" associated with high acoustic impudence were a quick way to spot gas reservoirs. Figure 6.10 shows the evolution of seismic attribute technology from the beginning (when the main use of seismic data was to identify major formation boundaries) to those developed in the early 2000 to assess the fluid type and fluid saturation.

Inversion is the process of finding an optimal model or set of models that fit the observation. The process generally involves some form of forward modeling and automated perturbation of the model parameters by minimizing

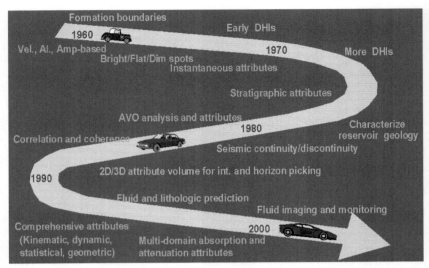

FIGURE 6.10 Evolution of seismic attribute technology (Aminzadeh, 2003). (For color version of this figure, the reader is referred to the online version of this chapter.)

an objective function. Inversion is a very loose term that sparks different associations with different geoscientists. Even within the seismic domain inversion can mean different things. In this chapter, inversion means the process of finding an impedance model for seismic trace data.

The input data determine the type of impedance that is inverted. Poststack seismic data are inverted to acoustic impedance, the product of density and P-wave velocity. Inversion of angle stacks yields elastic impedance, or effective impedance as it is sometimes referred to. Shear impedance, the product of shear velocity and density, can not only be inverted from stacked shear data but also be derived from prestack P-wave data. Simultaneous inversion of angle or offset stacks can yield volumes of acoustic impedance, shear impedance, and density. The latter is usually ill defined and is often constrained to the relationships observed in well logs (Pendrel, 2001).

In the following section, adapted from Selva et al. (2001), we describe many of the concepts referred to above in the form of a case history. This case history shows one of the conventional approaches for reservoir characterization using classical geostatistical approaches. Other approaches neural network, fuzzy logic, and genetic algorithms will be described later on.

6.4.7   A Case History to Illustrate Reservoir Characterization

This case history is adopted from a paper presented at the 7th International Congress of the Brazilian Geophysical Society: Selva et al. (2001). A horizon-based statistical analysis method was used to map a reservoir unit in Eastern

Venezuela. This resulted in about threefold increase in the original volumetric oil reserve estimates. Communication between four previously identified individual units became apparent leading to a single reservoir unit interpretation. The study involved a data set comprised of 23 wells and a 3D stacked migrated seismic data set covering approximately 80 km². The sedimentary column studied consists of fluvial to shallow marine deposits interpreted as part of the Miocene Foredeep Sequence, going from 3rd to 5th order sequences. The R4U reservoir is interpreted as a 5th order lowstand system tract.

Many statistically likely net-sand maps were generated by cokriging the RMS amplitude from seismic with net-sand values. An average sand distribution map was generated from 100 simulations with equal statistical weights. The net-sand map showed a clear E–W channel-shaped sand body that could not be identified previously using well data or seismic data alone. These results were used as an input to a proposal to drill three new horizontal wells in order to drain the upgraded recoverable reserves of 4.8 MMbbl.

6.4.8   Geologic Setting of Lobo Field

Lobo Field was discovered in 1952. Since then 23 vertical and 5 horizontal wells have been drilled. A total of 25 hydrocarbon-bearing reservoirs are present, with seven of them contributing to 90% of the original oil in place (OOIP) of 150 MMbbl. Reservoir units are between 5500′ and 6400′ in depth with the oil density ranging from 10° to 16° API. The consensus was that by drilling horizontal wells parallel to structure would allow production under very limited draw down. This in turn will lead to a limited deformation of the OWC, reducing water coning and sand production, thus increasing the life of the well. Also, because of the reservoir exposure with 1000 ft of horizontal sections, it was expected to increase the productivity index by a factor of five with respect a vertical well, this being the main justification for drilling horizontal wells. Decision to further develop the field with horizontal wells has led to major improvements in the performance. Consequently, mapping techniques presented here are of paramount importance to select future well locations.

The Lobo Field is located in the southern margin of the Eastern Venezuela Basin, in the Greater Oficina Area. This region is interpreted as the platform zone of the foreland basin. Is an extensional province with associated normal faulting trending N60°E. Locally, the trap is a homoclinal truncated by a normal fault, which provides the structural closure. Lateral closure is estimated as stratigraphic. Main reservoirs are located in the foot-wall block of the fault, showing a preferential NW–SE trend. Production becomes from Oficina and Merecure formations, developed in a complex fluvial-deltaic system. The units, from medium Miocene to Oligocene, interpreted as foredeep deposits, overly passive margin (Cretaceous) sedimentary rocks. Most reservoirs consist of interbeded sandstones and shales. Coal layers are common in the column.

The stratigraphic column of Lobo Field encompasses sediments of the first order passive margin and foredeep sequences. The passive margin sequence consists of interbeded Cretaceous sands and shales that overly metamorphic and igneous rocks of the Guyana Shield. On top, there is a major regional unconformity. The foredeep sequence comprises siliciclastic sediments ranging from Late Oligocene-Early Miocene to Recent. Two second-order cycles were identified in this sequence: a transgressive cycle that contains the Merecure, Oficina, and Freites formations; and a regressive cycle that include Las Piedras and Mesa formations. The Late Oligocene–Early Miocene to Medium Miocene is represented by the Merecure and Oficina formations, last conformed by interbeded deltaic to shallow marine sands and shales. For more details on the geologic setting and stratigraphy in this area, see Di Croce (1995) and Parnaud et al. (1995).

Detailed stratigraphy distinguished 10 third-order sequences within the foredeep cycle. Reservoir units represent fourth to fifth order sequences. Each sequence can be separated into stratigraphic system tracts. These high order sequences are from local tectonic pulses and sea-level oscillations that affected the foreland basin. The R4U reservoir consists of fluvial-meandering channel deposits that include a sandy bed-load channel facies and an overbank. The channel shows an SW–NE trend (Fig. 6.11).

The reservoir is composed of medium-grained, medium-sorted, and unconsolidated sandstones with porosities in the range of 21–27% and permeabilities from 300 to 1000 md. Average net pay thickness is 15 ft. Initial pressure of the reservoir is 2550 psi, with a permanent datum of 6070. Water drive is the main production mechanism. Previous interpretations assumed four individual reservoirs and placed four different oil–water contacts in them. On the basis of this new reservoir characterization study, it was determined that we have a single reservoir with a common oil–water contact estimated at 6200 ft. Figure 6.11 shows a display of sand distribution based on the RMS velocities and their correlations with sand thickness. More details to follow.

6.4.9 Methodology

To accomplish the task of reservoir characterization (generating the porosity and thickness maps), we followed a conventional geostatistical method in Fig. 6.12 show the flow diagram of the process that was used to generate final results. Data preparation and examination are very important to do quality control and to get a general feel for the available data and to understand the ranges of different parameters. Examination of data is facilitated by use of different statistical tools such as histograms, crossplots, matrix plots, and cross-correlation as shown at the bottom of Fig. 6.11. At the top section of Fig. 6.11 shows the

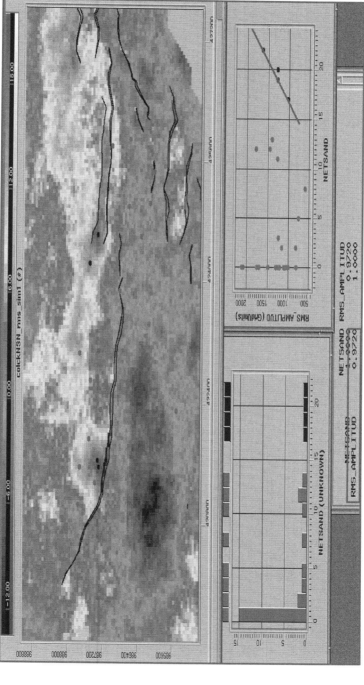

FIGURE 6.11 Histogram of the well control net sand and its correlation with RMS velocity net-sand thickness used to generate kriged net-sand distribution. (For color version of this figure, the reader is referred to the online version of this chapter.)

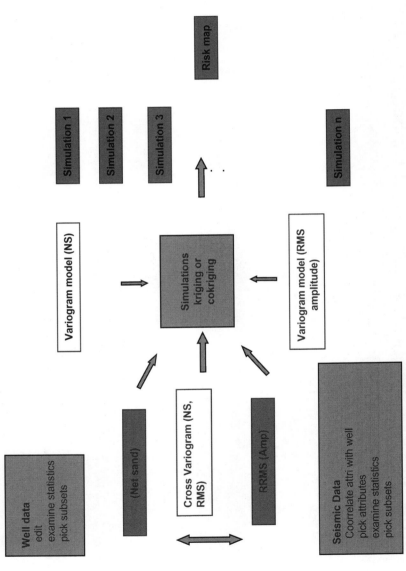

FIGURE 6.12 Flow diagram of reservoir characterization process. (For color version of this figure, the reader is referred to the online version of this chapter.)

TABLE 6.3 Step-by-Step Description of the Methodology

Step	Details and Comments
Data preparation	Data editing, choice of boundaries for mapping, and grid size
Data examination	Creation of histograms and matrix plots
Attribute choice	Compare seismic attributes and well properties correlations
Spatial statistics	Calculate variogram and correlogram
Kriging	Extrapolate well properties away from the well using well property variogram models
Kriging with external drift	Extrapolate well properties away from the well using variogram models and seismic at grid points as a guide
Colocated kriging or cokriging	The same as kriging except the seismic information usage is not limited to grid points only
Cokriging	Extrapolate well properties away from the well using variogram and cross-variogram models and seismic data
Simulation	Create multiple (100) realizations of cokriged results
Risk analysis and interpretation	Based on the simulation results, create the predicted value, and associate uncertainty at the proposed well location
Ground truth test	Test the prediction results against new wells and examine to ranges of predicted values and true drilling results

intermediate kriging result of net sand using well data alone. We examined different seismic attributes to constrain extrapolation of well data for kriging with external drift or for cokriging. This was accomplished by comparing the correlation coefficients of different seismic attributes (at well locations). RMS amplitude was selected as the most reliable attribute in establishing reasonable relationship with net sand and to a lesser extent with porosity.

Table 6.3 shows the major steps for such study. For technical details and other methods, see Deutsch and Journel (1998). It should be noted that to make full use of seismic attributes and including contributions from different attributes combination of different attributes using factor analysis or cluster analysis should be used to establish better correlations between seismic attributes and well properties.

6.4.10   Results and Conclusions

In this section, we provide several examples of the analysis and main conclusions of the study. As we note from Fig. 6.11, extrapolating well data only by kriging provides a map which shows general trends which lacks the details that seismic has to offer. As we outlined in Table 6.3 different kriging and cokriging methods can be used to include seismic information in extrapolating well information away from those locations. Among these methods are kriging with external drift. Colocated cokriging and cokriging of well data and seismic using the cross-variogram of the RMS amplitude and net-sand values as well as variograms of each data component.

All these methods can be applied to create any reservoir property. The confidence level on mapping results, however, is dependent on the well coverage and extent of correlation between the well properties and seismic attributes. Figure 6.13 shows porosity map generated from cokriging of porosity against RMS amplitude. The net-sand map showed a clear E–W channel-shaped sand body that could not be identified previously using well data or seismic data alone.

Given the fact that many other realizations of the net-sand map can be derived from the same data set, we generated a large number of simulations. A number of those simulation results are shown in Fig. 6.14. An average sand distribution map was to evaluate results and their level of uncertainties, we calculated the mean and standard deviations of net-sand distribution at a location that a new well was derived not used in the original data set for extrapolation. New prediction for several new well locations was made with the associated uncertainties. The yellow curve (to the left of Fig. 6.15a) shows the fitted normal distribution of the RMS amplitudes to the actual data points associated with the "clean" sands with over 7 ft thickness. The right hand side of Fig. 6.15a shows the crossplot of sand thickness (horizontal axis) against the RMS amplitudes. Figure 6.15b shows more details on the statistics of the RMS

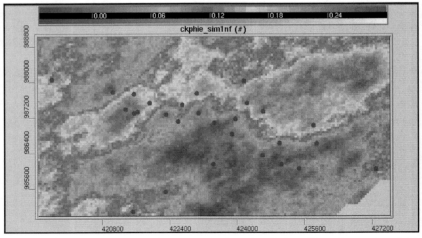

FIGURE 6.13 Cokriged of porosity using seismic and well data. (For color version of this figure, the reader is referred to the online version of this chapter.)

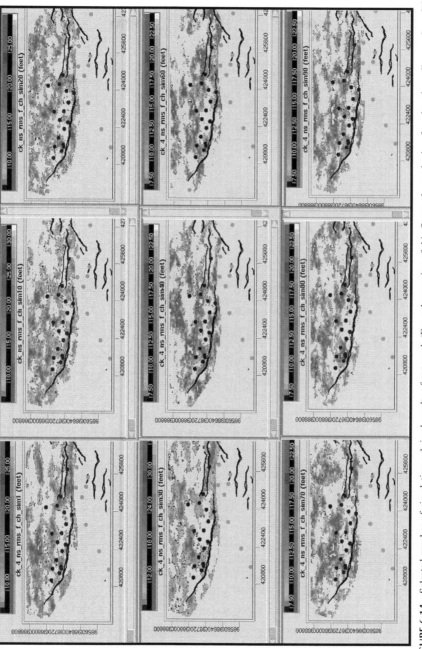

FIGURE 6.14 Selected number of simulation cokriged results of net sand. (For color version of this figure, the reader is referred to the online version of this chapter.)

A

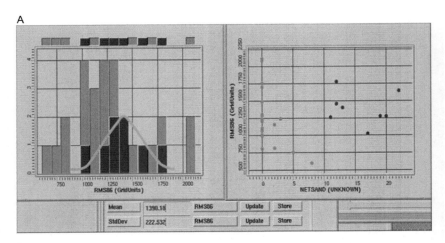

B

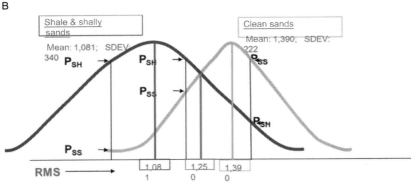

FIGURE 6.15 (a) Distribution of the sand thickness and (b) normal distribution of the RMS amplitude for both shale/shaly sand (left) and clean sand (right). (For interpretation of the references to color in this figure legend, the reader is referred to the online version of this chapter.)

amplitudes both for the clean send (right hand side and the shale and shaly sand the left hand side).

The normal distribution curves highlighted in Fig. 6.15b we could draw the following conclusions:

- For very high RMS amplitudes there is a good chance to have sand.
- For very low RMS amplitude there is a good chance to have shale.
- For RMS values to the right of the blue line (intersection of the two curves) it is more likely to have clean sand than shale or shaly sand.

the above conclusions assumed that they were in a good data area and withing the channel environment.

Figure 6.16 shows the location of the LG527 horizontal well at which the sand thickness was predicted to be 21 ft thick, with the standard deviation of

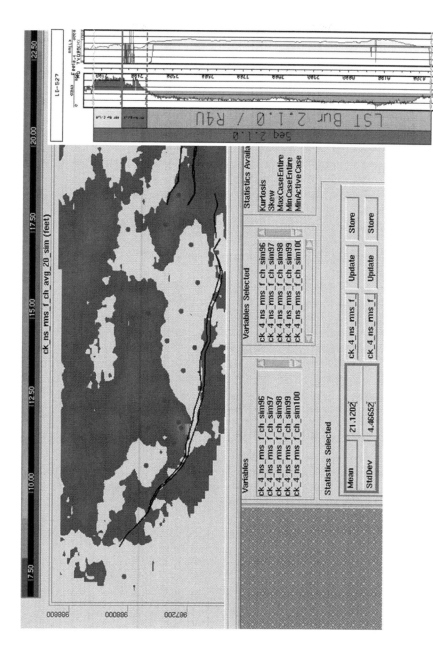

FIGURE 6.16 Predicted value of the sand thickness (bottom left) validated by the well logs (right). (For color version of this figure, the reader is referred to the online version of this chapter.)

4.4 (bottom left corner). The actual well showed a net-sand value of 25 (the log to the right) which was within one standard deviation of the predicted value. These results were used as an input to a proposal to drill three new horizontal wells in order to drain the newly upgraded recoverable reserves of 4.8 MMbbl.

6.4.11 Geophysics in Reserves Estimation

Among the objectives of reservoir characterization is the estimation of recoverable reserves. Up until 2010, this was primarily accomplished through the use of production data and well logs. Using these tools and methodologies engineers classify reserves as proven (greater than 90% chance), probable (greater than 50% chance), and possible (less than a 50 per). By using these metrics an appraisal of an oil and gas field can be performed for valuation purposes.

The SPE Oil and Gas Reserves Committee (OGRC) recently released *Guidelines for Application of the Petroleum Resources Management System* (PRMS). The new, 221-page document replaces the 2001 "Guidelines for Evaluation of Reserves and Resources" with expanded content that is updated to focus on using the 2007 PRMS to classify petroleum reserves and resources. For more details see OGRC-SPE (2011). Figure 6.17 shows key components of reserve assessment as defined by the Society of Petroleum Engineers and other cooperating professional societies who established the PRMS guidelines. Many of the terminologies in the PRMS of Fig. 6.17 are relatively new. More traditional terminologies and some common concepts associated with reserves are provided below with the respective definitions. For more details see the EIA (1998) document. Here are some highlights.

6.4.12 Original Hydrocarbon in Place

Reserves are those quantities of crude oil, natural gas, and natural gas liquids that are anticipated to be commercially recovered from known accumulations from a given date forward. Reserve estimates involve using different factors with varying degrees of uncertainty. The accuracy of those estimates depends largely on the amount of reliable available geological and engineering and geophysical data at the time of the estimate. The original hydrocarbon in place (OHIP) is expressed in terms of different factors in Eq. (6.1):

$$\text{OHIP} = 7758 \times \text{GRV} \times \text{NTG} \times \phi \times (1 - S_w)/\text{FVF} \qquad (6.1)$$

where GRV, NTG, ϕ, S_w, and FVF are gross reservoir volume, net to gross, porosity, water saturation, and formation volume factor, respectively. Among the factors impacting OHIP, geophysical data (3D seismic), are most effective in predicting GRV. Reservoir structure, its extent and thickness, are determined from seismic data with relatively high degree of confidence in many cases.

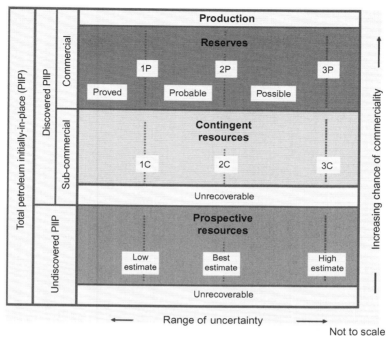

FIGURE 6.17 PRMS. © Society of Petroleum Engineers (SPE), World Petroleum Council (WPC), American Association of Petroleum Geologists (AAPG), Society of Petroleum Evaluation Engineers (SPEE). 2007. "SPE Petroleum Resources Management System (PRMS) Guide for Non-Technical Users". SPE.org. http://www.spe.org/industry/docs/PRMS_guide_non_tech.pdf (For color version of this figure, the reader is referred to the online version of this chapter.)

Determination of reservoir quality (a function of NTG, porosity, and facies) and fluid saturation, although could be helped by the use of advanced geophysical methods, would involve higher level of uncertainties.

The relative degree of uncertainty can be conveyed by broadly placing reserves into one of two categories—proved or unproved. Two basic methods are commonly used by industry to prepare reserve estimates—the deterministic and probabilistic methods. The deterministic method yields a single best estimate of reserves based on known geological, engineering, and economic data. The probabilistic method uses known geological, engineering, and economic data to generate a range of estimated reserve quantities and their associated probabilities. Each reserve classification gives an indication of the probability of recovery.

6.4.13 Proven Reserves

Proven reserves are those quantities of crude oil, natural gas, and natural gas liquids which geological and engineering data demonstrate with reasonable certainty to be recoverable in future years from known reservoirs under existing economic and operating conditions. Proven developed (PD) reserves include proved developed producing reserves and proved developed behind-pipe reserves. Proven developed producing reserves are only those reserves expected to be recovered from existing completion intervals in existing wells. PD behind-pipe reserves are those reserves expected to be recovered from existing wells where a relatively minor capital expenditure is required for recompletion. Proven undeveloped reserves are those reserves expected to be recovered from new wells on undrilled acreage or from existing wells where a relatively major expenditure is required for recompletion.

6.5   UNPROVEN RESERVES

Unproven reserves are considered less certain to be recovered than proven reserves. Estimates of unproven reserves are based on geologic and/or engineering data similar to that used to estimate proven reserves, but technical, contractual, economic considerations and/or Securities and Exchange Commission (SEC), state or other regulations preclude such reserves from being classified as proved. Unproved reserves may be further subclassified as probable and possible to denote progressively increasing uncertainty of recoverability.

6.5.1   Probable Reserves

Probable reserves are estimates of unproved reserves which analysis of geological and engineering data suggests are more likely than not to be recoverable. For estimates of probable reserves based on probabilistic methods, there should be at least a 50% probability that the quantities of reserves actually recoverable will equal or exceed the sum of the estimated proved plus probable reserves.

6.5.2   Possible Reserves

Possible reserves are estimates of unproved reserves which analysis of geological and engineering data suggests are less likely to be recovered than probable reserves. For estimates of possible reserves based on probabilistic methods, there should be at least a 10% probability that the quantities of reserves actually recovered will equal or exceed the sum of the estimated proved plus probable plus possible reserves.

Up until 2010, geophysics was not used in reserves estimation extensively, (Table 6.4) as it was considered a less than reliable tool. However, in December 2008, the U.S. SEC, published news rules regarding the determination of oil and gas reserves that, for the first time, consider seismology a "reliable technology" that can be used in this determination. The new rules become effective for reports issued after January 1, 2010. The new rules do not and cannot possibly specify all the specific technologies that are admissible. However, the new rules certainly open the door to the geophysical community in

TABLE 6.4 Reserve Estimation Techniques

Method	Comments
Volumetric	Applies to crude oil and natural gas reservoirs. Based on raw engineering and geologic data.
Material balance	Applies to crude oil and natural gas reservoirs. Is used in estimating reserves. Usually of more value in predicting reserves and reservoir performance.
Pressure decline	Applies to nonassociated and associated gas reservoirs. The method is a special case of material balance equation in the absence of water influx.
Production decline	Applies to crude oil and natural gas reservoirs during production decline (usually in the later stages of reservoir life).
Reservoir simulation	Applies to crude oil and natural gas reservoirs. Is used in estimating reserves. Usually of more value in predicting reservoir performance. Accuracy increases when matched with past pressure and production data.
Nominal	Applied to crude oil and natural gas reservoirs. Based on rule of thumb or analogy with another reservoir or reservoirs believed to be similar.

From EIA (1998).

several areas. The following are just a couple of examples of the applicable areas in geophysics.

First, the definition of fluid contact levels: in the current SEC rules, the definition of "lowest known hydrocarbon" or LKH and the "highest known hydrocarbon" or HKH requires production and flow test. With the introduction of reliable technology in the new rules, "flat spot" or other DHI technology, can potentially be used to establish LKH and/or HKH if it can be demonstrated that it shows hydrocarbon–water contact in the formation with reasonable certainty, consistency, and repeatability. This can have big impact on reserve booking. For example, if a trap is faulted and only one block has been penetrated and flow tested by a well, it may be difficult to book the other blocks under the current rules but possible with the "reliable technology" provision upon demonstration of reasonable certainty of the "flat spot" technology and transmission across faults.

Second, in regard to the permission of both deterministic and probabilistic estimates and disclosure of probable and possible reserves, the current rules only allow the disclosure of proved reserve estimates using deterministic method. The new rules also allow probabilistic estimates and the disclosure of probable and possible reserves in addition to proved reserves. Geophysical technologies in combination with other technologies can potentially make reserves estimation and booking possible without waiting for as many wells

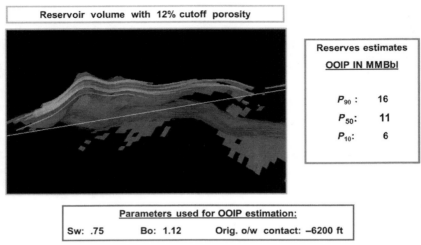

FIGURE 6.18 Original oil in place (OOIP) estimates with associated upper and lower limits, based on different input parameters. (For color version of this figure, the reader is referred to the online version of this chapter.)

and expensive flow tests as mandated by the current rules. The advances in geophysics provide alternative reliable technologies for reservoir parameter distributions. For example, seismic stratigraphic interpretation, reservoir porosity, and hydrocarbon saturation estimation have become routine practices in addition to traditional structural mapping. Many geophysical technologies, such as rock physics, seismic inversion, petrophysical inversion, and 4D seismic and history matching, provide important inputs to probabilistic reserves estimation.

Figure 6.18 demonstrates how geophysical information has been used for estimating reserves for the case history described earlier on Lobo Field (from Selva et al., 2001). A large number of simulated reservoir models after cokriging of the seismic data with the well data, as described earlier, also discussed in Chapter 5, yielded the upper and lower limits of the reservoir volumetrics or the OOIP. The reserves estimation was done using different cut offs for porosity and for 75% water saturation, 1.12 Barrel of oil factor and 6200 ft deep original oil/water contact. The mean value or P_{50} of the reserves is estimated at 11 million Barrels with P_{90} and P_{10} were estimated to be 16 and 6 MMbbl, respectively. The input parameters used for OOIP estimation are shown at the bottom of Fig. 6.18.

6.6 UNCONVENTIONAL RESOURCE ASSESSMENT

Resource/reserves estimation for unconventional resources has many similarities with those for the conventional resources but there are many similarities. Table 6.5 shows many parameters that are relevant to shale reservoir

TABLE 6.5 A Shale Gas Resource Assessment

Parameter	Low (P90)	Medium (P50)	High (P10)
Area (acres)	80	100	120
Thickness (ft)	300	600	900
Porosity (%)	4	6	8
Recovery factor (%)	10	20	40
Matrix gas saturation (%)	30	60	90
Gas storage capacity (scf/tonne)	10	30	60
Shale density, rho (g/cm^2)	2.2	2.4	2.6

resources assessment, many of which are similar to those of the conventional oil and gas reserves.

As indicated by Von Lunen et al. (2012), unconventional reservoirs also exhibit heterogeneity, and a failure in recognizing them can lead to economic grief. Oil sands reservoirs containing channel systems can have non-reservoir fills. Gas shales can contain non-reservoir facies that also act as frac barriers. In some shale gas plays, one should avoid faults and diagenetically enhanced open fracture zones, since such fracture zones become conduits for water to break through the seals bounding the resource container and consequently drowning the wellbore. Thus, accurate definition of the container is an important step in estimating the resources and reserves in unconventional reservoirs.

Since the reflection boundaries are not very well defined in such reservoirs, other inversion methods are used. Among useful geophysical tools for characterization of the fracture system in shale reservoirs are: curvature attributes, transverse anisotropy, coherency analysis (a measure of dissimilarity of the adjacent seismic traces), and edge detection algorithms. These tools are used in conjunction with borehole image log data from horizontal wells and core data. Elastic inversion methods can also help predict *in situ* shale gas resources and reservoir brittleness, trough calibration with the well log data. Estimates of SRV, proppant emplacement, and recovery factors are generated after comparison with production logs and decline curves.

Combining microseismic and seismic data have proven useful in characterizing shale reservoirs. (Maity and Aminzadeh, 2013). Usher (2012) also demonstrated usefulness of seismic and microseismic data integration in yielding good economic forecasts for such resources. It is pointed out that interpretive quantification does not always correlate well with production performance. For example, in some cases, significant amounts of the seismic energy created by pressure pumping are caused by releasing stress along a preexisting fault. In this case, the abundance of induced seismicity does not

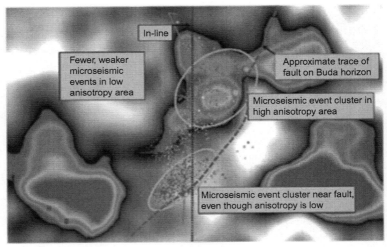

FIGURE 6.19 Contextual map of microseismic response and azimuthal anisotropy. (For color version of this figure, the reader is referred to the online version of this chapter.)

necessarily imply successful stimulation of reservoir rock. To demonstrate this, a relatively poor performing well, with low production rate, was selected from an Eagle Ford study where seismic and MEQ data were analyzed jointly. Figure 6.19 shows the frac intersected zones of high azimuthal anisotropy correspond to hypocenters, where the well crossed a large fault, implying abundance of hypocenters may have been caused by stress released along the fault.

Nevertheless, the frac may not have stimulated significant hydrocarbon production from the nearby formation, and the fault may also have acted as a thief zone for the frac fluid. He concludes adapting multivariate statistics techniques would allow more effective integration of the myriad seismic, petrophysical, and engineering attributes now available is changing the game, and is capturing the attention of the engineers. Specifically, he selected five variables: well bore length, Young's modulus (brittleness), azimuthal anisotropy (open fracture proxy), and both 10 and 32-Hz spectral decomposition.

While it makes intuitive sense that longer well bores provide more contact with the target formation, higher brittleness is an indicator of rock "fracability," and high velocity anisotropy suggests areas of open fractures. These attributes, individually, would not deliver the expected results. However, upon integration of the selected variables using nonlinear regression, a single attribute called Eagle Ford FracFactor was created. The higher values of this attribute had a strong correlation against maximum monthly production. Figure 6.20 shows the resulting model attribute was mapped to predict different zones of expected production, which can be used to high grade or phase the drilling and completions campaign.

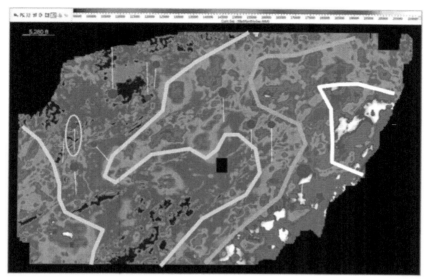

FIGURE 6.20   Eagle Ford predicted maximum monthly production zones. (For color version of this figure, the reader is referred to the online version of this chapter.)

Thus, it is important to predict the reservoir properties and the associated resource in place volumes before the frac program. We also want to determine the fracability, the existing faults and those to be avoided (including the potential concern for induced seismicity), carrying out geomechanical studies to determine the principal stress direction as well as characterization of natural fracture systems (orientation of natural fractures, their distribution, and whether they are open or closed). Some of this information should be collected both during and after the frac program, to determine its effectiveness and impact on the any major changes in the estimates of the resources. More details on this will be provided in Chapter 9.

As suggested by Von Lunen et al. (2012), it is important to evaluate the modification of the target reservoir into a state that permits economic production. This requires us to monitor the stimulated rock, identify bypassed resource pay, verify the resource confinement after stimulation, and predict or forecast the hydrocarbon delivery success from stimulation-induced changes in observed geophysical characteristics.

In shale gas exploitation, we must determine the extent of both the natural fracture system and the frac-induced fractures, the likelihood of achieving the desired SRV based on rock mechanical properties, the initial state of stress both vertically and laterally, and the effectiveness of seals and barriers to isolate the producing rock media. These data are provided or might be provided in the future, with improved analysis, by microseismic data. Figure 6.21 shows one of the steps toward resources evaluation where the volumetric

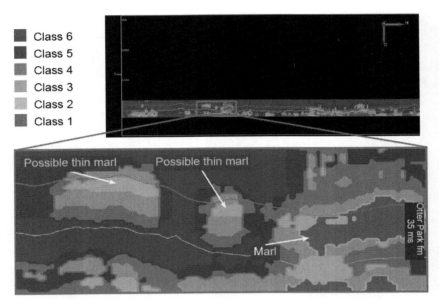

FIGURE 6.21 Volumetric segmentation of six classes of segmentation, using OpendTect software. *From Von Lunen et al. (2012). Courtesy of the Society of Exploration Geophysicists.* (For color version of this figure, the reader is referred to the online version of this chapter.)

segmentation of six different classes of segmentation where Marl reservoirs are classes 1 and 2 (with a lower confidence).

REFERENCES

Aminzadeh, F., 2003. http://www.geoinfo.com.ar/pdf/AMGE05-Met-Attributes_sigfrido.pdf.

Deutsch, C.W., Journel, A.G., 1998. GSLIB Geostatistical Software Library and User's Guide. Oxford University Press, Berlin, Heidelberg and Newyork.

Di Croce J., 1995. Eastern Venezuelan Basin: Sequence Stratigraphy and Structural Evolution, Ph.D. Thesis, Rice University.

EIA Document, 1998. Appendix G—Estimation of Reserves and Resources: http://www.eia.gov/pub/oil_gas/natural_gas/data_publications/crude_oil_natural_gas_reserves/historical/1998/pdf/appg.pdf.

Gaskari, M.R., Mohaghegh, S., 2007. Fieldwide reservoir characterization based on a new technique of production data analysis: verification under controlled environment. In: SPE Eastern Regional Meeting held in Lexington, Kentucky, U.S.A., October, pp. 17–19, SPE 11120.

Goodway W., Chen, T., Downton, J., 1997. Improved AVO fluid detection and lithology discrimination using Lamé parameters; lr, mr and l/m fluid stack from P and S inversions. CSEG National Convention Expanded Abstracts, pp. 148–151.

Maity, D., Aminzadeh, F., 2013. Fracture Characterization in Unconventional Reservoirs Using Active and Passive Seismic Data With Uncertainty Analysis Through Geostatistical Simulation, Annual Technical Conference and Exhibition, New Orleans, LA, September 28-October 2, 2013, SPE-166307-MS.

OGRC, SPE, 2011. PRMS Guidelines http://www.spe.org/spe-app/spe/industry/reserves/index.htm.

Parnaud, F., Pascual, J.C., Truskowsky, I., Gallango, O., Pasalacqua, H., Roure, F., 1995. Petroleum geology of the central part of the Eastern Venezuelan Basin, in Petroleum Basins of South America, AAPG Memoir 62.

Pendrel, J., 2001. Seismic inversion—the best tool for reservoir characterization. CSEG Recorder, January 2001.

Robertson, J.D., February 1989. Reservoir Management Using 3-D Seismic Data. Geophysics: The Leading Edge of Exploration 25–31. 16. Nolen-Hoeksema . . .

Schatzinger, R.A., Jordan, J.F., 1999. Reservoir Characterization, Recent Advances, AAPG Memoir 71.

Selva, C., Aminzadeh, F., Diaz, B.M., Porras, M.J., 2001. Using geostatistical techniques for mapping a reservoir in Eastern Venezuela. In: Proceedings of the 7th International Congress of the Brazilian Geophysical Society.

Usher, C.T., 2012. 3D Data Aid Shale-Field Development, American Oil and Gas Reporter, December 26, 2012.

von Lunen, E., Jensen, S., Leslie-Panek, J., 2012. Strategies in geophysics for estimation of unconventional resources. The Leading Edge 31, 1090.

Reservoir Monitoring

SUMMARY

As was discussed in Chapter 6, the goal of reservoir characterization is to use all the available data to create a model for the reservoir with as accurate estimates of the reservoir properties as possible. Accurate prediction of reservoir performance relies on the proper definition of the frame of the reservoir which is the rock matrix with empty pores. Reservoir characterization determines hydrocarbon distribution and the pathways or barriers impeding flow toward producer wells.

The key phrase here is "all the available data." Thus, as we produce from the reservoirs, new data become available. This includes the production data, updated decline curves, and possibly new seismic data. Creating an updated reservoir model or "dynamic model" is an important step to better understand any important changes in the reservoir characteristics. This information is crucial to do a better job in reservoir management and optimize production. It is also important when we need to make certain interventions such as enhanced oil recovery (EOR) (to increase permeability) and artificial lift (to increase pressure). It is also important to get updated information about the reservoir properties when we need to do an infill drilling. In short, we need to do an effective reservoir monitoring and surveillance during the producing life of a field and mapping of oil–water and gas–oil interfaces is necessary for understanding the fluid dynamics.

The implementation of geophysical tools in such monitoring has the potential to recover billions of barrels of bypassed oil and delay the abandonment of many marginal fields.

As the remaining hydrocarbons in the reservoir become more difficult to recover, fluid movement in the reservoir needs to be more closely monitored. The location of remaining hydrocarbons must be known in order to plan production wells and the injection schemes. Also, the manner in which injected fluids move and make contact with the target oil must be known in order to evaluate and, if necessary, correct the recovery project. Information on the preferred direction of fluid flow within the reservoir volume is imperative in the planning of production to achieve maximum volumetric sweep efficiency.

Engineers require high-resolution, numerical information about the spatial variation, vertically and laterally, of a reservoir's volumetric fluid-flow properties. The goal is to improve reservoir production and injection, increase production rate, optimize sweep efficiency, and avoid detrimental fluid movement—coning, channeling, and water breakthrough in producers. Reservoir properties are defined with high resolution at the wells from well logs and core analyses. They provide high-resolution, micro-, and macroscopic data at the immediate vicinity of the well. Interpolation of these data between wells, relating them to the interwell flow properties and scaling up the reservoir data over to the field size, is a difficult challenge. In a reservoir model spatial 3D information about reservoir and laboratory-scale variations and their distribution through the structure and reservoir architecture and textural heterogeneity and their relation to fluid-flow properties—porosity and permeability are required. Pressure-transient analysis and controlled core measurement well flow data provide the reservoir flow permeability and the preferred direction of flow. Geostatistical analysis can be used to combine aerially continuous but low-resolution seismic information and integrate with aerially sparse high-resolution, high precision log, and core data sampled over the reservoir volume. This is accomplished by creating models of reservoir property variations that are constructed to honor well log and core data as well as spatially calibrate the computed seismic attributes with well data.

In this chapter, we address three aspects of reservoir monitoring. They include time-lapse geophysics, EOR monitoring, and CO_2 sequestration and monitoring. Although there are other aspects of reservoir monitoring, the above three areas are deemed to be of the most importance, especially in the applicability of geophysical technologies to the reservoir monitoring issues. A few other related issues including monitoring the hydraulic fracturing process and monitoring unconventional reservoirs, especially in connection with changes in the fracture networks, are discussed in Chapter 9.

7.1   TIME-LAPSE GEOPHYSICS

Full knowledge of the reservoir properties and their subsequent updates would allow designing optimum development and production plan, strategically locating production and injection wells as well as infill drilling programs and sidetracks. The updated information would help make necessary revisions in the development plans, optimize reservoir management procedures, and enhance the value of new drilling campaigns. Knowledge of the frequently updated reservoir model would also help with forecasting arrival of water in a producing well or understanding the effectiveness of sweep from injected water or gas. Integral part of obtaining reliable updates about the reservoir is to collect "time-lapse" or episodic geophysical data including 4D seismic data, time-lapse electromagnetic data, time-lapse petrophysical data, and microearthquake data. Geophysical measurements are the only tools that can provide deterministic reservoir monitoring information between wells. The data could constrain the stochastic models and reduce uncertainty in predicting reservoir performance. Geophysical monitoring integrated with other data promises to provide valuable information about reservoir fluid movements and reservoir geological heterogeneities. All geophysical monitoring tools are based on measurements of physical properties and their contrasts over the production life of a reservoir. Various geophysical measurements are sensitive to different properties like elastic, electric, magnetic, and electromagnetic parameters, and they all have different resolutions. The properties measured must be calibrated with reservoir parameters obtained from independent measurements in wells.

Time-lapse or four-dimensional geophysical data are the repeated acquisition of seismic or other geophysical data at time intervals over the producing life of the reservoir to assess the impact of production or the results of water/gas injection for EOR, on the reservoir fluid saturation and pressure. The changes in geophysical parameters from repeated time-lapse measurements can be related to corresponding changes in reservoir properties like fluid pressure and saturation. The changes in parameters like seismic amplitude, velocity, V_p/V_s ratio, and other seismic attributes along with changes in gravity and EM measurements are applied in such studies. Such correlations would then be associated with other operational factors related to reservoir-driven mechanisms such as solution gas or water drive. Furthermore, we can obtain updated correlations with the corresponding field production data like production and/or injection rates and volumes, pressures in and around wells, and composition of produced fluids in wells.

Table 7.1 shows the time-lapse properties measured by various geophysical tools and techniques and the related reservoir properties inferred

TABLE 7.1 Application of geophysical tools for reservoir monitoring

Geophysical Technique	Physical Property Measured	Reservoir Property Inferred
Time-lapse 4D surface seismic	-Acoustic impedance volume -Seismic waveform changes -Seismic attributes, wave velocity change	Fluid saturation distribution. Fluid flow behavior. Reservoir dynamic changes of pressure and temperature
Time-lapse 4D VSP survey	-Acoustic impedance changes -Amplitude attenuation -Seismic wave anisotropy	Fluid saturation distribution. Fluid flow behavior. Reservoir dynamic changes of pressure and temperature
Cross-well seismic	-High-resolution tomographic imaging -Changes in acoustic impedance	Reservoir vertical conformance Fluid flow paths
Microseismic continuous monitoring (permanent surface and well bore)	-Microseismic events from shear slippage -Passive seismic waves from elastic failure due to stress field alterations -Hypocenters of induced microtremors from reservoir activities	Reservoir anisotropy Fluid flow pathways
Electromagnetic borehole and surface measurements and cross-well EM	-Time varying magnetic field induced electrical fields -E-amplitudes are proportional to impedance changes and hence resistivity	Direct inference of water saturation or flood front in a reservoir under water injection or active aquifer drive
Electroseismic (ES)	-Seismic from electro kinetic coupling changes -Resistivity changes	Fluid saturation changes with production and injection activities
Microgravity surface and borehole	-Minute gravitational field change due to water replacing oil or gas -Density differences	Fluid saturation changes with production and injection in reservoir
Satellite InSAR	-Remote measurement of surface deformation -Poroelastic relaxation	Reservoir volumetric change
Surface-buried tiltmeter	-Surface deformation -Poroelastic relaxation	Reservoir volumetric change

from the changes in physical properties. The same technique however, is not applicable in all fields.

7.2   4D SEISMIC

The 4D or time-lapse seismic method seeks to detect and characterize production-related changes in oil and gas reservoirs by recording seismic data at different times and measuring time-lapse changes in the seismic signal. Most reservoir changes occur because of changing pore pressure and/or fluid saturation levels during oil and gas production, but they may also arise from temperature and porosity changes within the reservoir, as well as from changes in the overburden, such as compaction or fluid movement along a fault or well bore. The time-lapse seismic method involves the recording of 1D, 2D, or 3D seismic data at several time steps; in each case, however, the technique is commonly referred to as the "4D seismic method," with the fourth dimension referring to calendar time.

4D seismic is now commonly used to locate bypassed hydrocarbons and permeability barriers, map water and steam fronts, monitor costly injectants, and detect potential CO_2 leaks. The method is invaluable for optimizing well locations and injection rates during field development and improving reservoir models for history matching and production forecasting. The first significant use of 4D occurred in the late 1980s, but since that time, it has gained in popularity to such a degree that many companies now require 4D acquisition and processing in their early field development plans. By the end of the last decade, the value of 4D was estimated to exceed US $4 billion in the North Sea alone, where it helped reduce drilling costs by more than 6% and contribute 5% additional oil reserves to each field (Amundsen and Landrø, 2007).

4D seismic works on the principle that changes in rock and fluid properties affect rock compressibility and shear strength, which, in turn, cause changes in the seismic response over time (Greaves and Fulp, 1987). Such changes are manifested as differences in seismic amplitude, phase, frequency content, velocity, travel time, and other physical attributes. In general, P-, or compressional, waves are sensitive to changes in both fluid content and pore pressure, while S-, or shear, waves are mostly sensitive to pressure (and relatively insensitive to fluid changes). Figure 7.1 illustrates these physical phenomena. By measuring both P- and S-wave time-lapse information, it is possible to invert separately for pressure and saturation changes, an area of great promise and active research.

Prior to acquiring 4D data, it is common to perform a feasibility analysis to determine whether 4D effects are detectable in the seismic data recorded over a particular reservoir. The 4D seismic method works best when seismic data quality is high and when reservoir rock and fluid properties are favorable. Good data quality implies that seismic signal to noise is sufficiently high to see a clear image of the reservoir and any oil–water or gas–oil contacts.

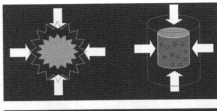

	ΔCompress	ΔShear
ΔSaturation	**+**	+
ΔPressure	**+**	**+**

FIGURE 7.1 Compressional and shear force on a sponge (yellow) and porous rock (purple). Top left, compressional force, sensitive to both fluid and pressure; bottom left, shear force, not sensitive to fluid, sensitive to pressure; and right, densitivity matrix. (For color version of this figure, the reader is referred to the online version of this chapter.)

Highly repeatable seismic acquisition and 4D coprocessed data (see below) also contribute to superior data quality. Favorable reservoir properties result from relatively shallow, compressible rock, such as young, unconsolidated sands, and from a high compressibility contrast between the *in situ* and displacing fluids, such as when water replaces high-GOR (gas–oil ratio). A feasibility study (Dasgupta, 2005) often includes an analysis of the value of information of the 4D project, which requires an estimate of the added value of 4D information for the purposes of well planning and reservoir management (Lumley, 2001), as well as the projected cost of acquiring such data multiple times throughout the life of the field.

Upon successful completion of a feasibility study—to be discussed in more detail later—confirming that 4D is favorable in a particular area, then seismic acquisition of the baseline and future monitor surveys must be planned. The goal of such a presurvey model study is to ensure that there is a good chance to detect changes in reservoir fluid and pressure seismically. Ideally, baseline and monitor surveys are acquired in an identical manner, with the same source and receiver configurations, time of year, source, and receiver type, so that subtraction of one data set from another reveals time-lapse differences that are related only to changes in earth properties. Moreover, the seismic data processing workflow should be identical for the two surveys. However, this is seldom possible and corrective measures must be taken in later data processing to make up for deficiencies in acquisition. 4D acquisition is especially challenging if the baseline survey was acquired many years prior to the first monitor survey. Many such 4D data sets exist with baseline and monitor surveys having significant differences in source and receiver type, source strength, offset and azimuth fold, shot and receiver spacing, shooting direction, etc. Perhaps the most important consideration in 4D acquisition is the minimization or elimination of nonrepeatable noise between surveys. Sources of nonrepeatable noise that can at least be partially mitigated include source and receiver positioning errors, differences in source or receiver response, and near-surface variations, such as tidal effects. Factors beyond one's control during acquisition include weather, unexpected overburden changes, and ambient noise.

Once 4D data are acquired, they must be processed in a manner that maximizes signal differences arising from changing earth properties (whether they are confined solely to the reservoir or also exist within the overburden) and minimizes extraneous differences not due to changes in the earth, such as nonrepeatable noise. There are generally three ways to process 4D seismic data. The first method simply processes each data vintage independently prior to subtraction. This has the advantage of simplicity but risks introducing artificial time-lapse differences that arise from differences in processing parameters or algorithms. In the second approach, data from each vintage are processed in parallel and use identical processing parameters, algorithms, and processing flows. This generally yields better results than the first method but may still generate artifacts because of differences in data quality or quantity between vintages. The third method, in which different vintages are processed simultaneously (or coprocessed) by mixing data at appropriate points in the processing flow, represents the best approach. Co-processing ensures greater consistency between vintages and enhances the statistical robustness of each step. More important, it allows the different data sets to be cross-equalized, so that time-lapse differences caused by changing earth properties are enhanced, while differences due to noise effects are reduced or eliminated. Cross-equalization can be performed either prestack or poststack and involves steps such as fold, offset, and azimuth equalization, global amplitude and frequency balancing, global and differential time and/or phase shifting, and data warping. Cross-equalization can be difficult to apply when overburden changes are present. Such changes are often considered noise when only reservoir changes are desired or expected, despite the fact that overburden changes may contain useful information about such important properties as well as bore stability and overpressure zones.

The processed 4D seismic data sets are usually interpreted or analyzed qualitatively, rather than quantitatively. Qualitative analysis often involves the use of one or more seismic attributes as a proxy to estimate the location, extent, and (perhaps) magnitude of change in a reservoir quantity, such as pore pressure and water saturation. For example, seismic amplitude differences between baseline and monitor surveys may reveal movement of the oil–water contact or indicate swept zones near an injection well. Forward seismic modeling is often used to attempt to duplicate or calibrate a time-lapse effect seen in field data. Seismic modeling is performed by first generating earth models and their associated seismic parameters, such as P- and

S-velocity and density, that correspond to each vintage of data. In 1D seismic modeling, seismic velocities and densities can be obtained from well logs recorded at different time steps. For 2D and 3D modeling, seismic parameters are usually derived from 3D static earth models, such as porosity and stratigraphic facies volumes, and dynamic models of pore pressure and fluid saturations obtained at each time step. In most cases, a rock physics model is used to generate seismic properties from reservoir earth properties. Qualitative 4D modeling and analysis have been used successfully for many years in a variety of areas as a means to locate swept or bypassed zones and improve the placement of sidetrack or infill wells.

Despite the benefits of a qualitative 4D analysis, it is often difficult to use in cases when both pressure and saturation change simultaneously, primarily because each of these properties can produce similar (and sometimes opposite) effects on seismic data. In order to invert for pressure and saturation changes separately and simultaneously, a quantitative analysis is usually performed by combining P- and S-wave information extracted from the 4D data, as mentioned above. Pressure-saturation inversion is an area of active research and a variety of approaches have been proposed over the years, most of which rely on a petrophysical model to establish the link between rock and fluid properties and their effect on seismic data (Batzle and Wang, 1992). In principle, other dynamic reservoir properties, such as temperature, can also be included in the inversion. In a sense, quantitative 4D inversion is an extension of the well-known amplitude-with-offset (AVO) inversion methods that have been used extensively by the industry for many years. In practice, however, 4D inversion methods are often difficult to implement within the time constraints of drilling schedules and field development plans, primarily because of the challenges involved in integrating a wide variety of core, log, seismic, and engineering data in an efficient and robust way.

In many fields, 4D seismic shows great potential in reservoir monitoring and management for mapping bypassed oil, monitoring fluid contacts and injection fronts, identifying pressure compartmentalization, and characterizing the fluid-flow properties of faults. This has a major impact on ultimate recovery and drilling efficiency, and it provides more accurate predictions of future reservoir production. As there are no changes in the reservoir geology, the differences in the data are attributed to dynamic changes in the fluid properties. 4D seismic technology provides information throughout the reservoir regarding fluid movements. Industry-wide market surveys identify time-lapse or 4D seismic technology.

Time-lapse seismic is now a proven technology for monitoring fluid movements and identifying undrained compartments in thick offshore clastic reservoirs. Several challenges still exist, however, in particular, the use of the technology for carbonate and thin-bedded clastic reservoirs (Amundsen and Landrø, 2007). It must be emphasized that some oil fields, for example, those with older and highly consolidate reservoir rocks or some carbonate reservoir or those with low GOR, may not be good candidates for 4D seismic monitoring. In such cases, other geophysical monitoring methods (e.g., gravity or electromagnetics based) approaches may be more applicable. In other situations, combination of different types time-lapse geophysical date may be desirable.

7.3 EVOLUTION OF 4D SEISMIC MONITORING

Since mid-1990s when 4D seismic projects began in North Sea, Indonesia, West Africa, and Gulf of Mexico, the technology has evolved and is now used routinely in field development projects. The technology has shown great economic value to reservoir and production engineers. Some of the 4D seismic projects showed excellent results, others with moderate and a few with no beneficial results. During the 1990s and the early 2000s, most producer and injector wells were vertical completions. The reservoir development planning required information between these wells for optimizing the number and location of the future wells. The development plans were updated once every 6 months. Any new information from production, injection and observation wells, and 4D seismic and petrophysical data was used in the updated development plan.

Now with the advent of horizontal and multilateral completions, and continuous monitoring in digital i-fields at the well bore, the needs for more quantitative data between wells have become more pronounced. In addition, smart wells with down hole control valves for allocating production from lateral have made it necessary for monitoring data available not just for planning but also for drilling and production operation optimization. Decisions to change production and injection allocation rates from individual laterals in multilateral wells are now based on the forecast of reservoir fluid flow. The water arrival at the well bore for different laterals will be predicted from this forecast and corrective action will be taken for opening downhole valves to choke or shut off production. The system thus provides a feed-forward control mechanism for proactive reservoir management solutions.

The life-of-field seismic project in Valhall field use permanent seismic sensors buried below the sea floor to acquire time-lapse seismic in a much higher frequencies (Barkved, 2012). The information is used not only for optimizing well production and adding infill wells, but also for geosteering during drilling operations of horizontal multilateral wells. In future, the data will need to be analyzed, interpreted, and integrated much more rapidly than it is done today. This will involve integrating continuous downhole measurements (e.g., reservoir pressure, temperature, flow rate, and water cut) with geophysical measurements (e.g., well logs, repeated seismic surveys, and microearthquake data). This will in turn provide an efficient field-wide reservoir model updating the mechanism on the basis of assessing the impact of production and injection on reservoir with time.

The next-generation time-lapse seismic probably would not be called 4D seismic. The reservoir monitoring would be for field-wide applications. New vision for geophysical reservoir monitoring techniques must include other geographic areas especially in the large oil fields of the Middle East, Caspian Basin. The technique must be applicable to reservoir with stiff carbonate rock frame and low GOR oil under water flood or CO_2 and water injection. They will be much more technically challenging and would update the reservoir models for prediction of performance in the interwell reservoir volume. Along with reflection seismic data, near continuous geophysical measurements like passive seismic, cross well seismic and EM, time-lapse EM, micro-gravity, etc., and computed attributes derived from them would be used for providing quantitative reservoir properties. These will be applied in the decision-making process for reservoir management production planning and by drilling engineers, and also for asset managers in volumetrics and reserves update. The process would improve the reservoir drainage, productivity, and infectivity indices and overall performance of the reservoir. The recovery factor would be enhanced and the expenses would be justified by the additional reserves included in the optimization process.

7.3.1 4D Seismic Technique

The changes are in oil, water, and gas saturation, injection fluid flood front location, and reservoir temperature and pressure. Following example from CGG Veritas shows repeated time-lapse seismic, a base survey for initial conditions in a reservoir and monitor survey after a later date and then the difference between them (Figure 7.2). The difference should indicate the changes in the fluid saturation and pressure from reservoir production at the end of lapsed time. The interpretation is usually qualitative. More recently, however, quantitative changes are being inferred from the data. Qualitative changes indicate where in the reservoir property changes took place. Quantitative changes

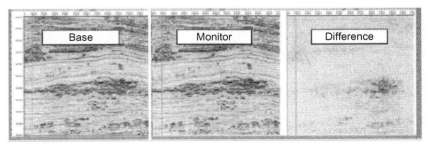

FIGURE 7.2 An example of 4D seismic with the base survey (left), monitor survey (middle), and the difference survey (right) from CGG Veritas. (For color version of this figure, the reader is referred to the online version of this chapter.)

define what changed and by how much. 4D seismic is being used for better understanding of reservoir fluid-flow paths, flow compartments, permeability barriers, fault seals, etc. This allows us to optimize field development plans, avoiding the costly mistake of drilling new wells into swept zones and finding new drilling opportunities. Many practical issues can complicate the simple underlying concept of a 4D project. These issues are practical questions about the fluid property changes.

As was shown in Figure 7.1, the physical basis for 4D seismic is rock physics. This relates the seismic time-lapse signal to reservoir pressure, saturation, and temperature. Changes in rock and fluid and other properties that affect compressibility and shear strength are measured by differences seismic wave front amplitude and energy, wave propagation velocity and travel time, phase, frequency, and acoustic impedance.

As seismic waves propagate through the reservoir deform rocks by compressing and shearing them. Fluid saturation changes in the reservoir affect the seismic compressional or P-waves, while changes in reservoir pressure impact both P and shear or S-waves. The results of 4D interpretation provide volumetric changes in reservoir properties between the wells. Successful 4D projects in North Sea and West Africa have contributed to dramatically improving forecasts of how a reservoir behaves during production.

Figure 7.3 shows two slices from 4D seismic data volumes from Statoil's Gullfaks field in North Sea. This example is from Solheim et al. (1999) where one of the early practical successful applications of 4D seismic for reservoir monitoring was demonstrated. The baseline seismic data were acquired in 1985 and the first repeated acquisition was in 1999. The map based on 4D

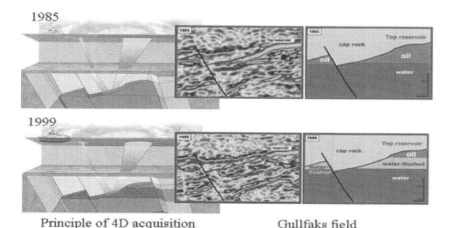

FIGURE 7.3   An example of changes in the seismic response after 15 years of production in Gullfaks field. *Courtesy of StatOil ASA. Reprinted with permission.* (For color version of this figure, the reader is referred to the online version of this chapter.)

seismic results proved to be more accurate than that derived from history-matched reservoir simulation model. 4D seismic signature indicated oil saturation in clastic Tarbert reservoir lying between faulted compartments. Subsequent drilling proved the bypassed pockets of undrained or bypassed oil between fault blocks. Since the first success, several 4D seismic monitor data have continued to be recorded in Gullfaks field. Due to the elastic nature of clastic Tarbert reservoir, the time-lapse signal is high. The mud-rich continental margin of North Sea, with highly elastic clastic rocks, usually demonstrates high time-lapse signal. In this environment, successful time-lapse studies were carried out during the 1990s. Early time-lapse studies were carried out during the 1990s.

The changes in seismic reflection amplitude between the two surveys are believed to be due to a significant depletion of the oil produced 1985 through 1999 (Figure 7.3). The amplitude difference of the top of the reservoir is related not only to reduction in oil saturation but also to the original oil-column height. As water replaces oil, the amplitude of the respective seismic reflection is substantially decreased; creating a "dimming effect" on what was a strong reflection from the top of the reservoir. The strong oil–water contact-related seismic response in 1985 has also been dimmed, due to production of oil. The smaller oil accumulation, to the left of the fault, was drained by 1999, whereas much of the oil was still to be recovered from the larger trap to the right of the fault.

In general, 4D seismic plays a key role, which is to maximize the economic return from oil and gas fields. The economic benefit in 4D seismic is derived from operational excellence and capital efficiency. Time-lapse seismic technology has become best practice in many organizations; it is used to monitor reservoir production, reduce reservoir uncertainties, and identify additional reserves. Effective use of technology can improve operational efficiency by means of fewer, better-placed, and safer wells (Figure 7.4).

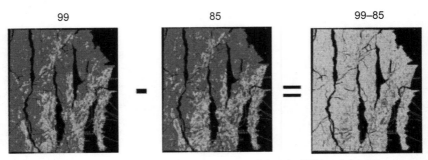

FIGURE 7.4 Seismic time lapse from different vintages (left and middle) and the difference section (right). *Courtesy of StatOil ASA. Reprinted with permission.* (For color version of this figure, the reader is referred to the online version of this chapter.)

Most 4D seismic analysis methods, including the two examples shown earlier, rely on the observation and interpretation of the changes in the seismic amplitude. However, as we discussed in Chapter 3, there are many other seismic attributes that change due to the changes in the reservoir fluid and other reservoir properties. They include seismic frequency, seismic phase, and others. Oldenziel et al. (2002) show how other seismic attributes or a combination of them can be used in observing changes in reservoir properties more effectively. The conventional amplitude differencing approach is shown in Figure 7.5A. Using a number of different seismic attributes, a time-lapse difference vector is created to assess the changes in such vector in a time-lapse data set. Figure 7.5B shows such changes. Comparing this against Figure 7.5A, we notice more clearly the locations where such changes have taken place (appropriately color coded).

Figure 7.5C result is based on the use of multiattribute technique, except for the fact that instead of simple differencing of the attribute vectors, artificial neural network (ANN) is used where known changes in the reservoir (based on production and well log data) are used as training set to the ANN

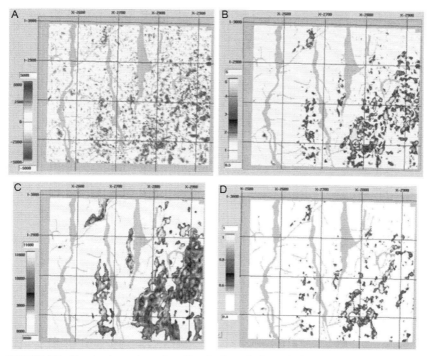

FIGURE 7.5   Results from alternative 4D seismic analysis methods: (a) Conventional seismic amplitude difference method, (b) multiattribute differencing scheme, (c) 4D object method, and (d) 4D meta-attribute approach. *Courtesy of dGBes.com* (For color version of this figure, the reader is referred to the online version of this chapter.)

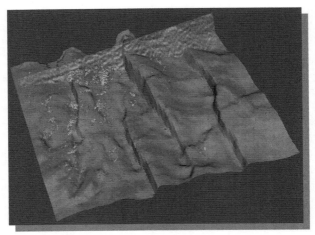

FIGURE 7.6 4D meta-attribute volume. *Courtesy of dGBes.com.* (For color version of this figure, the reader is referred to the online version of this chapter.)

to highlight other changes in the seismic attribute vector set. Finally, Figure 7.5D shows a more refined version of Figure 7.5C called the meta-attribute output where additional information and an interpretive-based approach were used to filter out unlikely reservoir-related changes in the output.

Figure 7.6 shows the 3D volume, highlighting 4D meta-attribute that goes beyond the simple traditional amplitude differencing concept.

7.3.2   Challenges in Linking Time Lapse

Seismic data provide structural and static information for a reservoir such as lateral extent of the reservoir, thickness, faults, porosity, among other reservoir properties. Time-lapse seismic acquired at different times measures changes in reservoir state. Consistency of the acquisition and processing of 4D seismic is a challenge. In most cases, the base line or initial seismic survey was acquired without plans for any future 4D seismic. The techniques used in acquiring the legacy seismic data is obsolete compared to the technology available at the time of new survey.

Table 7.2, from Oldenziel (2003), defines the challenges associated with linking 4D seismic to dynamic reservoir models. The challenges are categorized into two categories. The second is to fully integrate 4D seismic data with reservoir engineering. Aside from the feasibility study discussed earlier, time-lapse seismic project for a field usually needs to be justified based on cost and potential gain. Once the project is approved, the work begins systematically. A 4D seismic workflow was published by Shell (Kawar et al., 2003) that has the following stages: feasibility study, acquisition design, data processing,

TABLE 7.2 Some of the Challenges of 4D-Seismic Technology
Implementation

Link 4D Seismic to Reservoir Properties	Integrate 4D Seismic with Reservoir Engineering
• Repeatability—acquisition, reprocessing, cross-equalization	• Integration—huge amounts of disparate data sets at different scales
• Interpretation—rock physics, quantitative	• Accelerate integration loop to increase benefit of data
• Lack of calibration data— validation of different methods	• Parameterization
• Decoupling of reservoir properties—pressure and saturation	• Nonuniqueness
• Definition of 4D attributes	• Automated history matching—misfit function, optimization algorithm, stopping criteria

data interpretation, and feedback loop of the results to the geological and dynamic reservoir models.

The updated models developed are used for planning field development and locating additional production and injector wells.

7.3.3   Feasibility Study of 4D Project

Feasibility study is the first stage of 4D seismic monitoring workflow. It is usually conducted for the given reservoir using measured rock, fluid properties, and reservoir simulation model. The properties are obtained from borehole data in core samples, well logs, well tests, and laboratory measurement of fluid properties. Detailed reservoir characterization is done using seismic and all borehole data reservoir: (1) a geologic model based on the initial baseline seismic data; (2) static and dynamic properties of reservoir rocks and fluids measured from cores, geophysical logs, and well tests; and (3) detailed field production and pressure data. The resulting reservoir model constitutes the best possible input to determine whether subsequent 4D monitoring is feasible or not (Khan et al., 2000; Vidal et al., 2000). The risks associated with a 4D seismic project include false anomalies caused by artifacts of time-lapse seismic acquisition and processing and the ambiguity of seismic interpretation in trying to relate time-lapse changes in seismic data to changes in saturation, pressure, temperature, or rock properties. The risk factors for 4D seismic must be carefully evaluated and modeled during the feasibility study (Meyer, 2001). A case history on oil field at the end of this chapter further highlights the feasibility study, modeling, and other aspects of a typical 4D project.

It should be noted that not all reservoirs are good candidates for 4D seismic monitoring. The rock and fluid proprieties must be favorable for the technique to work. The reservoir rock frame must be compressible, with high matrix porosity; the fluids should have high compressibility, for example, oil with high gas–oil ratio and injected water or gas reservoirs with water drive. About 10 years ago, Lumley (2004) forecasted that the expansion of the 4D market in different parts of the world was uneven: North Sea (80%), offshore West Africa (7%), offshore North America (6%), and Far East (4%). Although this has changed to some extent, still Northwest Europe continues to utilize time-lapse seismic extensively. The Middle East, Offshore Brazil, Offshore West Africa (Angola and Nigeria), and Offshore North America (Gulf of Mexico) are becoming new growth areas for 4D seismic. Approximately 500 seismic time-lapse repeat surveys have been conducted, so far in 250 oil and gas fields worldwide.

7.4　GRAVITY DATA FOR FLUID MONITORING

While 4D seismic has proved to be an effective reservoir monitoring tool, as indicated earlier, some reservoirs may not be suitable for it. Other geophysical tools may be more effective. For example, field-wide gravity monitoring offers a unique possibility to directly measure changes in mass as they take place in a producing reservoir. The bulk density change in a reservoir due to changes in pore fluid density can generate a change in gravity that can be measured on the surface with precision gravity meters. The difference in density between oil, gas, and water is relatively, small however, over the large volume in the porespaces in the reservoir they can cause changes in gravity of >100 μgal and can be detected down to 5 μgal (Bate, 2005). Such measurements provide an useful tool for identifying the movements of gas/fluid contacts, optimizing production, and estimating in-place reserves. The measurements can be made in time-lapse episodes or monitored continuously using permanent sensors over the area of interest. The precision in gravity monitoring far exceeds those required for gravity surveys for hydrocarbon exploration.

Four time-lapse gravity surveys have been acquired over the Troll field in order to image and monitor the rise of the liquid contact during gas production. In the Sleipner field in the North Sea, two surveys show the average change in density when CO_2 is injected in water-filled sandstone. Such results have later been confirmed by well and seismic data and are actively used for reservoir management and simulation purposes.

Since the gravitational anomaly is small, its detection above the normal survey noise level can be improved with a gradient-type survey. In a gravity gradient survey made at the same recording station, two gravimeters separated by a short distance vertically, simultaneously, record the acceleration due to gravity. The two gravity measurements are provided by instruments that are matched and aligned to a high level of accuracy. The noise as well as other gravitational corrections will be the same for both readings since they are acquired at the same location and time. The gravity anomaly is, however, sensed at two different elevations, so its vertical derivative is obtained. This derivative may be integrated to obtain the small-density change that results from fluid saturation change. Gravity gradiometers measure the spatial derivatives of the gravity vector. The most frequently used and intuitive component is the vertical gravity gradient, G_{zz}, which represents the rate of change of vertical gravity (g_z) with height (z). Equation (7.1) shows the corresponding relationship:

$$G_{zz} = \frac{\partial g_z}{\partial z} \approx \frac{g_z\left(z + \frac{l}{2}\right) - g_z\left(z - \frac{l}{2}\right)}{l} \qquad (7.1)$$

Gravity gradiometry offers key advantages over other techniques. Superconducting gravity gradiometer incorporates superconducting circuits which can be balanced such that its responses to gravity gradients are largely independent of all linear and angular accelerations applied to the instrument. It has low noise, negligible scale factor drift, and optimum mechanical stability.

7.5   GRAVITY METHOD FOR EOR MONITORING

The change in a rock's pore fluid content from oil or gas to water leads to a change in average reservoir density because high-density water replaces low-density hydrocarbon fluids. These small-density changes in the reservoir have effect on the gravitational acceleration that can be measured at the surface. Usually, a large area of the reservoir is affected by this fluid replacement. In the Prudhoe Bay Field, Alaska, surface-gravity surveillance of a gas-cap water flood was implemented. The objective is to maintain pressure in the gas cap, while the oil production is in declining phase. It would have been prohibitively expensive to drilling numerous surveillance wells to monitor water movement adequately. The gravity technique measured the changes in gravitational field. The results indicated that density changes associated with water replacing gas are readily detected by use of high-resolution surface-gravity measurements. From the data interpretation of reservoir pressure and the compaction of the reservoir layers, overproduction time are obtained. This is necessary for reservoir and production engineers for prediction of reservoir performance and for updating reservoir simulation models. Compaction of the reservoir is caused due to the withdrawal of hydrocarbons during production. This also results in subsidence of the overburden layers above the reservoir and in a decrease of reservoir porosity. The replacement of oil and gas with water causes mass addition or a positive gravitational field anomaly. Inversion of time-difference gravity field changes generates a model of reservoir density changes and hence the fluid saturation changes. The decrease in reservoir porosity and the removal of hydrocarbons should have a corresponding effect on seismic response, in terms of acoustic impedance, amplitude, as well as frequency content, and this can be identified and mapped in the time-lapse seismic.

7.6 MONITORING CO$_2$ SEQUESTRATION AND EOR MONITORING

Geophysical technologies, based on repeated seismic or other data mesuarements, have also been used for monitoring the effectiveness of CO$_2$ injection for enhanced oil recovery. This is usually preceded by monitoring an effective CO$_2$ capture and sequestration program. While we will not go into any details, the concept is similar to what we discussed earlier on waterflooding or production impact on the rocks and how seismic response can be impacted by it. For an overview of the time-lapse seismic monitoring of CO$_2$ sequestration and EOR operation, see Wilson and Monea (2004), Friedman (2007), Lumley (2010), and Verdon (2012).

A low gas–oil ratio in the residual oil offers the most favorable conditions for seismic monitoring of CO$_2$. This is the case for the Weyburn field in Saskatchewan, Canada. This field has emerged as a test bed for evaluation of different time-lapse seismic and microearthquake CO$_2$ and water monitoring programs (Li, 2003; Verdon, 2012; Wilson and Monea, 2004).

Weyburn is a mature carbonate field with thin-fractured carbonate reservoir. Its pool of 1.4 billion-barrel reserves has produced for 48 years, mostly through waterflooding. To reverse the production decline, as shown in Figure 7.7, different approaches, including vertical and horizontal infill drilling, have been implemented. More recently, CO$_2$ flooding has been implemented with the expectations of a more significant production increase.

Li (2003) reports on the use of 4D seismic monitoring of CO$_2$ flooding Weyburn field. The objective is to get a better understanding of the changes in the reservoir and the status of CO$_2$ fluids in the reservoir zones. Specifically, the goal is to separate the effects of reservoir pressure, saturation, and

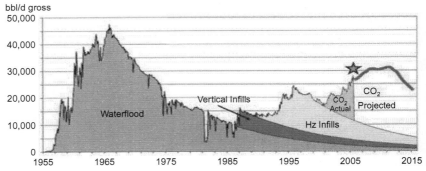

FIGURE 7.7 Evolution of EOR processes in Weyburn field. *Courtesy of Weyburn-Midale two project.* (For color version of this figure, the reader is referred to the online version of this chapter.)

fracturing-induced azimuthal anisotropy. This is accomplished by rock physics modeling and 4D AVO and Lambda–Mu–Rho inversion in conjunction with 4D converted P–S wave analyses. It is also contemplated that three-component 4D VSP analysis can further highlight fracture influences. See Chapter 3 to revisit some of these geophysical methods.

Figure 7.8 compares the 4D anomaly map with production engineering data such as up-to-date cumulative injection volume, hydrocarbon pore volume, and CO_2 recycle ratio. We notice that the higher injection volumes correspond impressively to strong 4D anomalies, for example, at injection patterns 3, 4, 5, 6, 9, and 11, where the HCPV (hydrocarbon pore volume) on average is at or above 10%. Pattern 12, which exhibits an off-trend seismic anomaly, also has a high injection volume but shows an extremely high recycle ratio (40%). In the horizontal producers south of pattern 12, abnormal GOR over 1200 was observed. Clearly, CO_2 breakthrough, probably introduced by NW–SE-oriented fractures, had occurred.

Other well-established, large-scale CO_2 injection projects involving monitoring are Sleipner in the Norwegian North Sea (Arts et al., 2004) and In Salah in Algeria (Riddiford et al., 2005). Figure 7.9 shows the entire process of CO_2 or Carbon, Capture and Sequestration, also referred to as CCS. To facilitate the process, one may use a software package such as CO_2-PENS software, developed by Los Alamos National Laboratory. CO_2-PENS links together physics-based

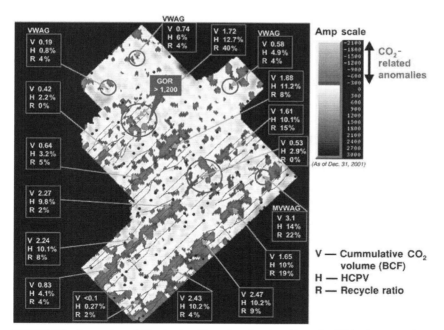

FIGURE 7.8   Integration of 4D seismic with engineering data. *From Li (2003). Courtesy of the Society of Exploration Geophysicists.* (For color version of this figure, the reader is referred to the online version of this chapter.)

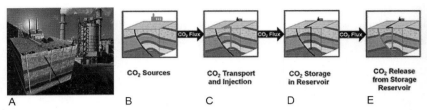

FIGURE 7.9 CO_2 capture (from the power plant or geologic formations, A and B), injection for EOR (C), Storage (D) and Release (E), from http://co2-pens.lanl.gov/. (For color version of this figure, the reader is referred to the online version of this chapter.)

process-level modules that describe the entire CO_2 sequestration pathway, starting from capture at a power plant and following CO_2 through pipelines to the injection site and into the reservoir. After injection, simulation of CO_2 migration continues through the subsurface where it may mineralize, dissolve into brine, or react with wellbore casing or cement. CO_2 may leak from the reservoir along wellbores or faults that lead back towards overlying aquifers or the surface. The model can be used to quickly screen sequestration sites or to perform a more detailed site-specific evaluation. Figure 7.10A from Lumley et al., 2008 shows a comparison of seismic time-lapse difference model with and without CO_2 and the actual difference data (left panel) against the actual 4D seismic difference data at Sleipner (right panel). This highlights the fact that contamination by several imaging artifacts caused by complex internal scattering and mode conversion and multiple interferences can complicate the interpretation. Nevertheless, there is a remarkable similarity of the model and the real data where the pointers are located.

While the opportunities for seismic and micro seismic monitoring of CO_2 are abundant, the challenges as maintained by Lumley (2010) should be addressed. They include:

1. aggregate impact of (a) the injection pressure, (b) saturation changes in the reservoir, and (c) the possible presence of multiple phases of CO_2 is challenging to separate;
2. complications can arise to evaluate pressure-saturation effects because of CO_2 reactive effects on the rock matrix or pore fluids;
3. rock physics and fluid analysis show that seismic compressibility is only weakly sensitive to CO_2 saturation levels beyond about 30% S CO_2 when present as a supercritical "fluid," and beyond about 10% S CO_2 at a supercritical "gas" situation;
4. estimating the density effect of injected CO_2 from seismic data is difficult except in the presence of very large amounts of CO_2 and the availability of high-resolution seismic data with low SNR and wide reflection angles;
5. highly nonlinear and nonunique nature of the seismic response to CO_2 makes it difficult to extract accurate information (such as travel time, velocity, and amplitude).

Figure 7.10B shows the time-lapse data at successive years of CO_2 flooding in Sleineper field.

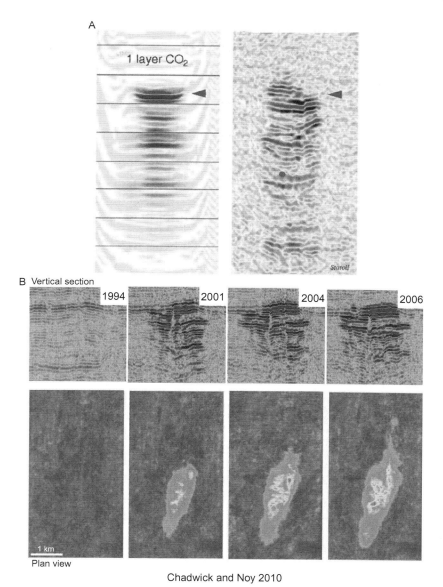

FIGURE 7.10   (a) Synthetic seismic model difference (left) image actual 4D seismic difference Image at Sleipner (right) *Lumley et al., 2008.* (b) Time-lapse seismic lines at Sleipner showing growth of the CO_2 plume from 1994 to 2006. *Courtesy of the Society of Exploration Geophysicists* (For color version of this figure, the reader is referred to the online version of this chapter.)

7.7   A 4D SEISMIC CASE HISTORY FROM GHAWAR OIL FIELD

This case history is adopted from Dasgupta (2005). It shows a time-lapse monitoring case project in Ghawar, world's largest oil field producing oil from Upper Jurassic Arab-D carbonate reservoir. Monitoring the advancement of floodfront from water injection in the Arab-D reservoir is a major challenge in carbonate rocks. Fluid flow anisotropy in the reservoir mapped over time represents a challenge for geophysical methods. The Arab-D reservoir gross thickness is about 100 m. The matrix comprises carbonate rock sequence of grainstones, packstones, and wacke stones. The original sedimentary textures have been altered in many parts of the reservoir by leaching, recrystallization, cementation, dolomitization, and fracturing. Reservoir is capped by anhydrite layers in Arab-C–D interval and is underlain by Jubaila formation, a sequence of tight fine-grained limestone and lime mud (Al-Husseini, 1997). Porosity ranges in the reservoir from less than 10% at the base to over 30% at the top. The permeabilities range from a few millidarcies to over 1500 millidarcies. The reservoir production is supported with peripheral water injection at the structural flanks for pressure maintenance. The reservoir waterflood front generally advances uniformly and sweeps the oil from the east and west flanks to the central crest of the structure (Figure 7.11).

The reservoir is well connected both laterally and vertically. However, in the mid-field region of Ghawar field, some producing wells have experienced premature water encroachment. This renders the flood front movement uneven in these localized areas. The inhomogeneities in reservoir properties are due to stratigraphic complexity caused by deposition, diagenesis, and sub-seismic faults and fractures (Mitchell et al., 2000). These have contributed to production anomalies like early water encroachment in producers. Monitoring of flood front between wells would provide "an early warning system" to the reservoir engineers. Timely availability of this information could prevent premature water breakthrough in production wells, identify untapped oil pockets in the reservoir, and increase the ultimate recovery of oil.

The study area has been under production for over 50 years and is now reaching maturity. Reservoir pressure is maintained by peripheral water injection, which is the primary driving mechanism for oil production. The faults and fractures often act as conduits for water encroachment. The Arab-D is a strongly undersaturated oil reservoir with no free gas. Also the injected water maintains the reservoir pressure above the bubble point which prevents the formation of a secondary exsolution gas cap. The bubble point pressure in the reservoir is about 10 MPa. Oil production in the reservoir is replaced with water with similar elastic moduli and the lack of free gas and relatively low GOR are expected to cause low 4D sensitivity. Table 7.3 shows the Arab-D reservoir fluid properties at the dry production area, at the oil water contact, and at the injector wells. With the reservoir becoming progressively mature, anomalous production from early water encroachment creates production challenges.

FIGURE 7.11   Ghawar field in Saudi Arabia Feasibility study area is indicated on the map. *Courtesy of Saudi Aramco.* (For color version of this figure, the reader is referred to the online version of this chapter.)

TABLE 7.3 Reservoir Fluid Properties in the Study Area

Fluid Properties	Sw (%)	Salinity (ppm)	Specific Gravity	GOR (Scf/ BBL)	Pressure BHFP (MPA)	Temperature (Degree)
Wells in dry area	14	215,000 TDS	0.75 oil 1.15 brine	510	15.1	90
Wells at floodfront	67	60,000 TDS	1.15 brine		22	90
Injection wells	85	45,000 TDS	1.05	510	31	90

Sw, water saturation; GOR, gas–oil ratio; SCFlbbl, standard cu feet per barrel; BHFP, bottomhole flowing pressure.

7.8 SEISMIC MONITORING CHALLENGE IN ARAB-D

Because of low impedance contrast between oil and water in the reservoir, conventional 4D seismic is unlikely to detect flood front changes, especially with the low annual depletion rate in the Arab-D reservoir. The key factors that dictate time-lapse seismic application for reservoir monitoring are sensitivity and repeatability. When the sensitivity of seismic time lapse or 4D attribute is low, the repeatability issue becomes more critical.

7.9 FEASIBILITY STUDY

A feasibility study consisting of rock physics petroacoustic data analysis and seismic modeling was conducted over a mature area in the field. The objective was to study seismic monitoring methods using permanent downhole sensors. Permanent sensors are expected to substantially improve repeatability of seismic measurements. Modeling quantified the seismic response of reservoir saturation changes as brine replaces oil. Seismic forward modeling was performed using Gassmann's fluid substitution equations. Reservoir fluid properties were measured and derived from a history-matched flow simulation model for each time step of production life—Arab-D reservoir oil production through the predicted total depletion. Reservoir static frame and rock grain properties were measured in the laboratory and inferred from well log data for 24 wells over the study area. Fluid substitution modeling concluded that because of low impedance contrast between the oil and injected brine and the stiff Arab-D carbonate reservoir frame, time-lapse surface 3D seismic or conventional 4D seismic is unlikely to detect the flood front within the repeatability of surface seismic measurement.

7.10 METHODOLOGY

The study included review of different seismic monitoring methods, both active seismic and passive seismic measurements. A history-matched fluid simulation model initialized from the beginning of production in 1953 and prediction run to reservoir depletion at a 1-year step. The model consists of 53 cells in X and 12 cells in Y direction with 18 layers. Properties in the simulation model included those of reservoir frame are porosity, x,y,z, permeability, and dolomite.

Volume and reservoir dynamic properties are water and oil saturation for every year from the initiation of production to reservoir depletion. Dolomite volume is related to the volume of solid phase. The model was tested for consistency with the well data at each of the 24 wells in the study. The impact of fluid saturation changes on seismic response was computed by forward modeling.

7.10.1 Petroacoustic Modeling

Well logs and core analysis data from 24 wells in the study area were used to build the rock physics and petrophysical models which relate reservoir properties (porosity, lithology, fluid saturation) to the elastic rock properties that impact the seismic response (bulk modulus, shear modulus, density). This provided elastic properties in each cell of the dynamic model (Figure 7.12). A linear relationship was used to compute density as a function of porosity, fluid saturation, and lithology grain density (Schön, 2004). Separate relationships were used for computing density for limestone and dolomite matrix. Dry rock matrix bulk modulus, as a function of porosity and lithology, was

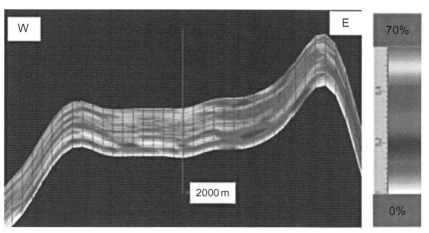

FIGURE 7.12 Simulation model of water saturation change 1953–2032 (80 years) Sw 73% change. *Courtesy of Saudi Aramco.* (For color version of this figure, the reader is referred to the online version of this chapter.)

modeled using regression analysis. Gassmann's equations were used to model velocity as a function of water saturation. From the density and velocity, acoustic impedance was computed in each cell. Synthetic time-lapse seismograms were computed from reservoir flow simulation model.

7.10.2 Seismic Forward Modeling

Seismic modeling quantified changes in seismic response as oil is replaced by water in the reservoir. Several simulation scenarios were tested to analyze the sensitivity of the seismic response to fluid substitution. The reservoir model and water saturation changes were obtained from the history-matched reservoir simulation model. For each 1-year time-lapse step of the dynamic simulation model, a corresponding elastic model and its seismic response for each cell were computed.

A layered overburden model was included above the reservoir in order to convert depth to time. Seismic response was computed trace by trace from the impedance grid converted to time. Reflection coefficients for each impedance trace corresponding to the vertical limits of the reservoir were computed and convolved with a zero-phase wavelet estimated from the 3D seismic data in the area. Changes in seismic response produced the synthetic time-lapse seismic attribute data that would quantitatively relate to changes in fluid saturation. At each time step, a 3D zero-offset seismic response was computed. The seismic response calculated for trace by trace was convolved with zero-phase wavelet estimated from the processed seismic volume in the area. The time lapse or 4D effect was quantified from attribute difference between base and monitor case. This assessed the sensitivity of the seismic response to changes in the reservoir due to production. Table 7.4 is a summary of the seismic-response sensitivity over production time.

TABLE 7.4 Time-Lapse Effect for Different Time Intervals

Time Lapse Period	Maximum 4D Seismic Attribute Change (%)	Seismic Time Lapse Attribute at a Study Well (%)
1953–2032 from initiation of production to expected depletion	6	4.9
2004–2032	3	2.1
2003–2004	0.4	0.3

The results of modeling have been used to select the most suitable seismic monitoring scenarios and to design the acquisition layout for a field pilot. Fluid substitution modeling in the 18-layer dynamic model, history matched with production data, indicates that the maximum change in water from initiation of production to reservoir depletion is 73.4%, and the corresponding change in acoustic impedance is 4.2% (Figure 7.13).

The low time-lapse sensitivity is exacerbated by the low depletion rate in Arab-D. Table 7.5 shows a comparison of 4D effects between Ghawar field and Gullfaks field in the North Sea. Time-lapse seismic has been extremely

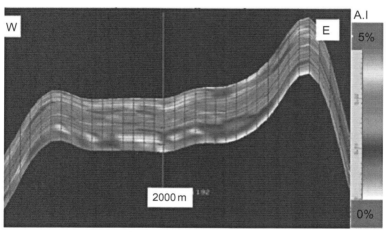

FIGURE 7.13   Acoustic impedance changes in the simulation model 1953–2032 4.2% change. *Courtesy of Saudi Aramco.* (For color version of this figure, the reader is referred to the online version of this chapter.)

TABLE 7.5 Comparison Between Ghawar and Gullfaks Fields Seismic Time-Lapse Effects

Reservoir Properties	Ghawar Field (Arab-D reservoir)	Gullfaks Field (Tarbert reservoir)
Porosity maximum	30%	35%
Dry rock bulk modulus	12–20 GPa (carbonate rock)	5–10 GPa (sandstone)
Water saturation change	60–70%	40–50%
Fluid compressibility change	80–100%	150–250%
Predicted acoustic impedance change (to depletion)	4–6% (over 80 years)	10–12% (over 15 years)

TABLE 7.6 Qualitative HSE Risk Comparison of Four Types of 4D Surveys

	Towed	OBC	Node	Permanent
Man-hours/200 km^2 survey	36,000	235,000	72,000	10,000
Major risk	Towing	Cable lay	Node lay	Gun work
CO$_2$ emissions per survey	High	High	Medium	Low

Source: Adopted from Barkved (2012).

successful and rewarding in Gullfaks. The predicted impedance change due to reservoir depletion in Arab-D is less than half of that in Tarbert reservoir. While this change takes over 15 years in Gullfaks, for Ghawar the change is over 80 years due to its low depletion rate.

Modeling shows that near-surface velocity variations over Ghawar by about 3% would override this time-lapse signal related to reservoir fluid replacement. Changes in near-surface conditions are the most challenging issue in repeatability of time-lapse seismic in Arab-D reservoir over Ghawar. It is difficult to discriminate between seismic response due to reservoir saturation changes with near-surface changes in velocities, depth of water table, and changes in coupling conditions. For 1-year production period (2003–2004), change in water saturation (Sw) generates a change in acoustic impedance of 0.4%. This small change is not detectable by time-lapse seismic. A time-lapse period of at least 5–6 years production would be required to produce the minimum detectable seismic difference due to fluid replacement.

The high rigidity of limestone–dolomite reservoir rock matrix and a small contrast between the elastic properties of pore fluids—oil and water are responsible for the weak 4D seismic effect from oil production. A feasibility study was conducted to quantify the 4D seismic response of reservoir saturation changes as brine replaced oil. The study consisted of analyzing reservoir rock physics, petroacoustic data, and seismic modeling. Seismic model of flow simulation using fluid substitution concluded that time-lapse surface seismic or conventional 4D seismic is unlikely to detect the flood front within the repeatability of surface seismic measurement.

7.11 PERMANENT RESERVOIR MONITORING

The technology for monitoring reservoirs for different applications is advancing in the data acquisition systems, sensors and data analysis. Permanent reservoir monitoring (PRM) systems are considered advantageous over its processors 4D seismic recording methods of "towed," OBC, and Node-based systems described earlier. Unlike towed streamer surveys, ocean bottom cable surveys, or node surveys, permanent sensors installed on the seafloor minimize the impact on existing oil field infrastructure and enable highly repeatable, cost-effective 4D seismic imaging in and around obstructed zones.

PRM achieves recording of high-resolution recording of seismic data from the floating production storage and offloading or platform or remote field control room. The seabed arrays of PRM with links to production facility, as shown in Figure 7.14, allow operators to affordably monitor production and injection performance on demand. This operation increases the recovery factor with lower drilling cost and improves EOR performance. As described in Bett (2013), operators can benefit from the innovative seabed seismic technology of PRM in the form of lower lifetime cost of operation.

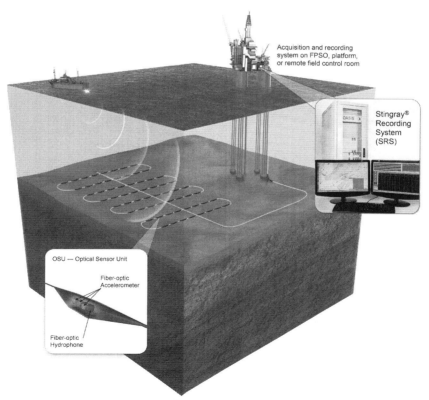

FIGURE 7.14 A Fiber-optic Stingray® Permanent Reservoir Monitoring (PRM) system, *courtesy of TGS-Norpec Company* (www.tgs.com). © TGS-NOPEC Geophysical Company ASA. *From Bett (2012)*. (For color version of this figure, the reader is referred to the online version of this chapter.)

In addition, the data acquisition process using the PRM system is faster, with reduced environmental risk. Table 7.6, adopted from Barkved (2012), makes the comparison on survey time, major risks, and the CO_2 emission level.

Van Gestel et al. (2008) show the results of the use of PRM in creating amplitude difference maps for the first nine surveys on Valhall. For a movie depicting the amplitude (reservoir) changes with time, see http://www.eage.org/images/cms/l_movie.gif. They are all relative to the first survey. The red lines mark horizontal well trajectories with perforations, and the black lines are interpreted faults. For other recent developments, see www.rmc.usc.edu, highlighting the ongoing research work of USC Reservoir Monitoring Consortium.

REFERENCES

Al-Husseini, M.I., 1997. Jurassic sequence stratigraphy of the western and southern Arabian Gulf. GeoArabia 2 (4), 361–382.

Amundsen, L., Landrø, M., 2007. 4D Seismic—Status and Future Challenges, GEO ExPro October 2007.

Arts, R., Eiken, O., Chadwick, A., Zweigel, P., van der Meer, L., Zinszner, B., 2004. Monitoring of CO2 injected at Sleipner using time-lapse seismic data. Energy 29, 1383–1392.

Barkved, O.I., 2012. Seismic surveillance for reservoir delivery from a practitioner's point of view. Education Tour Series, EAGE.

Bate, D., 2005. 4D reservoir volumetrics: a case study over the Izaute gas storage facility. First Break 23, 69–70.

Batzle, M., Wang, Z., 1992. Seismic properties of pore fluids. Geophysics 57, 1396–1408.

Bett, M., 2012. The Value of PRM in enabling high payback IOR, PESA News Resources. Dec 2012/Jan 2013, p. 45.

Bett, M., 2013. The value of PRM in enabling high payback IOR. PESA Resources, No. 121, December/January 2013.

Chadwick, R.A., Noy, D.J., 2010. History – matching flow simulations and time-lapse seismic data from the Sleipner CO_2 plume. In: Vining, B.A., Pickering, S.C. (Eds.), Petroleum Geology: From Mature Basins to New Frontiers–Proceedings of the 7th Petroleum Geology Conference. Petroleum Geology Conferences Ltd. Published by the Geological Society, London, pp. 1171–1182.

Dasgupta, S.N., 2005. When 4D seismic is not applicable: alternative monitoring scenarios for the Arab-D reservoir in the Ghawar Field. Geophys. Prospect. 53, 215–227.

Friedman, S.J., 2007. Geological carbon dioxide sequestration. Elements 3, 179–184.

Greaves, R.J., Fulp, T., 1987. Three-dimensional seismic monitoring of an enhanced oil recovery process. Geophysics 52, 1175–1187.

Kawar, R., Hatchell, P., Calvert R., Khan, M., 2003. The workflow for 4D seismic. Middle East Oil Show, 9–12 June 2003, Bahrain.

Khan, M., Waggoner, J.R., Hughes, J.K., 2000. 4D cause and effect: what do reservoir fluid changes look like on seismic? SPE Paper 63132, 2000 SPE Ann. Tech. Conf., 10pp.

Li, G., 2003. 4D seismic monitoring of CO_2 flood in a thin fractured carbonate reservoir. Leading Edge 22, 690–695.

Lumley, D.E., 2001. Time-lapse seismic reservoir monitoring. Geophysics 66, 50.53.

Lumley, D.E., 2004. Business and technology challenges for 4D seismic reservoir monitoring. The Leading Edge 23 (11), 1166–1168; SEG Tulsa.

Lumley, D., 2010. 4D seismic monitoring of CO_2 sequestration. Leading Edge 29, 150–155.

Lumley, D., Adams, D., Wright, R., Markus, D., Cole, S., 2008. Seismic monitoring of CO_2 geo-sequestration: realistic capabilities and limitations. SEG Conference Expanded Abstracts, 2841–2845.

Meyer, R., 2001. Time-lapse reservoir geology. CREWES Res. Rep. 13, 905.

Mitchell, J.C., Lehmann, P.J., Cantrell, D.L., Al-Jallal, I.A., Al-Thagafy, M.A.R., Mjaaland, S., Wulff, A.-M., Causse, E., Nyhavn, F. SPE Technical Conference 2000. Integrating Seismic Monitoring and Intelligent Wells. Paper SPE 62878, SPE Annual Technical.

Oldenziel, T., May 2003. Time-lapse Seismic within Reservoir Engineering, PhD Thesis, Department of Applied Earth Sciences, Delft University of Technology.

Oldenziel, T., Meldahl, P., Ligtenberg, H., Digranes, P., Stronen, L.K., 2002. Multi-attribute analysis of 4D anomalies using pattern recognition technology. In: SEG Annual Meeting, Expanded abstract, Salt Lake City.

Riddiford, F., Wright, I., Espie, T., Torqui, A., 2005. Monitoring geological storage. In: Wilson, M., Morris, T., Gale, J., Thambimuthu, K. (Eds.), Salah Gas CO_2 Storage Project. Greenhouse Gas Control Technologies, Proceedings from the 7th Greenhouse Gas Control Technologies Conference. vol. 2. Elsevier, pp. 1353–1359.

Schön, J.H., 2004. Physical Properties of Rocks, Volume 8: Fundamentals and Principles of Petrophysics Handbook of Petroleum Exploration and Production.

Solheim, O.A., Hilde, E., Ekren, B.O., Strønen, L.K., 1999. The Gullfaks 4D seismic study. Petrol. Geosci. 5, 213–226.

van Gestel, J., Best, K.D., Barkved, O.I., Kommedal, J.H., 2008. Integrating frequent time-lapse data into the reservoir simulation modelling of the Valhall field. In: 70th EAGE Conference and Exhibition, Rome, Italy.

Verdon, J. P., 2012. Microseismic monitoring and geomechanical modelling of CO_2 storage 11.

Vidal, S., Longuemare, P., Huguet, F., 2000. Integrating geomechanics and geophysics for reservoir seismic monitoring feasibility studies: SPE Paper 65157, 2000 SPE Ann. Tech. Conf., 10pp; vol. 19(1), January 2001, pp. 24–45.

Wilson, M., Monea, M. (Eds.), 2004. IEA Greenhouse Gas Weyburn CO2 Monitoring & Storage Project Summary Report 2000–2004, 273pp.

Geophysics in Drilling

The process of drilling an oil or gas well requires knowledge of all geologic features expected to be encountered along the way—from the surface of the ground to the target reservoir. Thus, in addition to steering the well so as to intersect hydrocarbon-bearing reservoirs, the reservoir engineer must assure to a reasonable degree of confidence that the well drills successfully and safely to the target. Geophysical measurements help ensure a successful drilling program. 3D seismic provides a picture of the subsurface from the surface to the target.

The subsurface image-based on geophysical techniques provide useful information to:

1. identify drilling hazards that may lead to an uncontrollable well;

Developments in Petroleum Science, Vol. 60. http://dx.doi.org/10.1016/B978-0-444-50662-7.00008-1

2. assist in site surveys for well construction and platform stability;
3. predict what lies ahead of and around the drill bit; and
4. illuminate what exists above and below the wellbore in a horizontal or highly deviated well.

Drilling hazards fall into three main categories:

1. lost circulation, or loss of drilling mud into the formation; and
2. abnormal formation pressures;
3. shallow gas pockets.

A pressure imbalance between the wellbore and the formation characterizes both the hazard conditions. Such imbalance, if severe, may lead to an uncontrollable well, even to a disastrous blowout.

8.1 SEISMIC DELINEATION OF STRUCTURAL DISRUPTIONS: FAULTS, FRACTURES, AND SINKHOLES

The geologic features that cause lost circulation during drilling may be imaged by seismic and other geophysical methods. Severe loss of drilling mud occurs when the wellbore intersects a loss zone. These zones are characterized by highly porous and permeable formations formed by either shattering the rocks or dissolving them. When the drill bit encounters a cavern and drills into the void, the entire weight of the drill string will have to be borne by the derrick, a condition for which it is not designed. Many blowouts along the Golden Lane of Mexico, in offshore Gulf of Mexico, have resulted from derrick collapse due to drilling into subsurface caverns.

3D seismic data provides a three-dimensional view of these geologic features so that they can be viewed from different perspectives—in either cross section or plan view. In order to be resolved by seismic data, structural disruptions—such as faults, fractures, or sinkholes—must have sufficient vertical displacement and lateral aperture or extent. Theoretically, the minimum seismic resolution is equal to about $\lambda/8$, where $\lambda =$ seismic wavelength $= V/f$, in which V is the seismic velocity and f is the dominant frequency. For typical carbonate rocks having an average seismic velocity of 20,000 ft/s and a dominant frequency of 30 Hz (cycles/s), the minimum resolution is about 80 ft. Thus, a fault must have a vertical throw of at least 80 ft in order to be discernible as such in seismic data. Similarly, a sinkhole must have collapsed about this much so that its edges may be identified.

Even when these features are below seismic resolution, and therefore, cannot be directly imaged, their presence may still be inferred by processing the data for a variety of seismic attributes. The presence of a small fault, fracture, or sinkhole, affects the amplitude of the reflected seismic energy. Such structural features scatter the incoming seismic wave in various directions and the resulting image has reduced amplitude. When viewed in plan view, or on time slices, the linear trends formed by these dim zones allow the seismic interpreter to delineate the locations and distribution of these sub-seismic faults or fractures. However, their vertical displacements, if any, cannot be determined from the seismic data.

In addition to diminished amplitudes, small faults and fractures often also manifest themselves as slight changes in seismic wavelet shape that are not easily discernible by eye. These small wavelet variations may be enhanced by a seismic attributes like semblance that measure similarity between adjacent traces. The similarity between adjacent seismic traces is computed by taking the semblance or coherence between them. If the traces are similar, the coherence is high, if they are unlike, their semblance is low. Coherency is used for identification and mapping of channel edges, reefs, faults, and fracture systems in 3D volumes. Coherence attributes highlight displays as dark colors zones of low semblance (Fig. 8.1 shows semblance attribute processing). Using this technique, a network of small faults or fractures may be mapped, despite our inability to pick them directly on seismic sections (Fig. 8.2).

Various indicators of structural disruption—displaced seismic reflection, reduced amplitude, low trace-to-trace semblance, structural dip, and

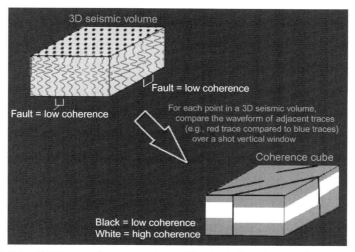

FIGURE 8.1 Multitrace attribute processing on 3D seismic for coherence shows areas of low coherence or semblance between traces as dark and high coherence are white. *Courtesy: http://kgs.ku.edu (Kansas Geological Survey).* (For color version of this figure, the reader is referred to the online version of this chapter.)

FIGURE 8.2   Coherence attribute processed 3D seismic data showing subsurface geologic features imaged with this technique. *Courtesy: Arcis/TGS*. (For color version of this figure, the reader is referred to the online version of this chapter.)

curvature—in many cases become redundant since they describe the same phenomenon. This is true, for example, when a fault is large and resolvable. In such cases, it is sufficient to map them from picking and mapping the seismic reflection alone or in combination with coherence processing. Where these faults and fractures are below seismic resolution, they become "sub-seismic," and we need to resort to further structural analysis using the other attributes.

8.2   HIGHLY POROUS FORMATIONS

Besides shattered rock, ultra-high-porosity, non-hydrocarbon-bearing rocks, such as dolomitized carbonate rocks, may pose lost circulation problems, albeit of probably less severity than fractures. The presence of these highly porous zones may be determined in the same manner as the porosity characterization of the actual hydrocarbon reservoirs that comprise the target of drilling, through interpretation of the acoustic impedance volume obtained from amplitude inversion of the seismic data.

As with the reservoir, the porosity of the problem formation typically varies linearly and inversely as the acoustic impedance. Therefore, a map of the average impedance of this formation should be able to delineate likely areas of high porosity, and hence, of potential lost circulation. Better yet, if several wells have been drilled and logged through this formation, an actual porosity–impedance relationship for this formation may be defined to transform the seismic impedance into predicted porosity. If, furthermore, drilling experience in this area can define a threshold porosity beyond which lost circulation becomes a serious concern, then the actual problem zones within this formation may be delineated in map form or in three dimensions.

8.3   MECHANICALLY WEAK ROCK LAYERS

There are cases where mechanically weak rocks, such as chalk, are interlayered with normal, competent rocks. If the mud weight called for in the drilling program exceeds the tensile strength of this weak layer, drilling will fracture the rock and the mud will be lost to the weak formation. The mechanical strength of a rock is best described by its Poisson's ratio, which depends on the ratio of the rock's compressional, V_p, and shear, V_s, velocities. Weak rocks have lower shear velocities than competent rocks; hence, their V_p/V_s ratios are higher.

Since seismic data acquired for reservoir characterization is typically compressional in nature, the resulting stacked-migrated seismic volume that the geophysicist interprets contains no direct information on the shear acoustic properties of the rocks. However, in its original, pre-stack form, the seismic data contain indirect measure of the rock's shear properties. The theory of seismic wave propagation states that, beyond vertical incidence, the reflected amplitude of an incident compressional wave also depends on the shear velocities of the rocks.

By measuring the variation of a reflection's amplitude with offset (AVO), an indirect measure of the shear velocity contrast at the top of this weak rock layer is obtained. The greater this velocity contrast, the larger is the AVO effect. Thus, in principle, weak rock layers may be analyzed in this manner. In practice, however, the analysis also depends on the fluid content of the rock and since the processing involved is complex (particularly for 3D seismic data), it is seldom done. The best practical option is to rely on lessons learned from past drilling experience through this particular weak layer.

8.4   SUBSURFACE CAVERNS

Subsurface caverns present a drilling risk, presenting a unique problem to seismic imaging. They have several characteristics which handicap imaging:

a. They are typically shallow;
b. They are three-dimensional in shape, not flat-lying;
c. Since they are voids, they have a large acoustic impedance contrast with the enclosing rocks; and
d. They are not the primary target of investigation.

Since near-surface caverns are not the primary objective, the seismic survey is not designed to image them. Because they are at shallow depths, the seismic fold (or multiplicity) of the data reflected from them is low and lacks large shot-receiver offsets. Hence, there exist few traces to stack at their depth. Their large impedance contrast generates strong reflections, while their shape scatters the impingent seismic energy. All these contribute to difficulty in focusing their images during seismic migration processing. Worse, they adversely affect the imaging of the underlying reservoir target.

In the Arabian Peninsula, open caverns occur so close to the surface that oftentimes their roofs breach the surface. The entire axial spine of the giant Ghawar Field anticline is dotted with numerous collapsed caverns (karst) and open caverns, that the resulting seismic image of the principal Arab-D carbonate oil reservoir and the deeper Khuff and pre-Khuff clastic gas reservoirs appear incoherent and uncertain. Figure 8.3 shows an example of an open cavern on the Arabian Peninsula and its expression on seismic data indicated with arrows. Such caverns when they are buried deeper pose drilling hazards in wells. These are predicted from 3D seismic data, ahead of the drill bit.

When a drilling project critically requires an accurate image of the shallow subsurface, a separate seismic survey, using high-resolution, low-energy, and low-effort seismic techniques, such as those applied for civil engineering and groundwater work, may be conducted. Instead of a heavy

FIGURE 8.3 Shallow collapse feature and its effect on seismic reflection data (arrow). *Courtesy: Saudi Aramco (Personal communication)*. (For color version of this figure, the reader is referred to the online version of this chapter.)

vibrator truck as a seismic source, small percussion or explosive sources are used. Shotgun blasts, blasting cap explosions, heavy weights dropped from a small height are some of these sources of seismic energy. Instead of hundreds or thousands of geophone receivers, a few dozen receivers suffice. The reduced hardware requirements and scope of operations require vastly reduced manpower to conduct the survey. However, the data acquired undergo the same processing as that for the deep seismic survey because the experiment remains the same and the goal is similar, to acquire an image of the subsurface.

As in all aspects of reservoir engineering work, data relevant to a specific objective (in this case, drilling hazards) from all sources should be considered before a course of action is taken. The structural images, lithologic predictions, and other inferences from geophysics should complement the thorough analysis of geologic data and previous drilling experience in order to fully illuminate the hazard problem. As a tool that provides abundant, but indirect, measurements of the regions not actually sampled by existing wellbores, geophysics contributes significantly to the understanding of drilling hazards, but only when properly integrated with all other data.

8.5 GEOPHYSICAL PREDICTION OF OVERPRESSURED ZONES

Overpressure or geopressure is a drilling hazard that is responsible for many well blowouts. These occur when the pore fluid pressure significantly exceeds that predicted from the normal compaction of sediments with depth. It becomes a hazard because, if not anticipated and prepared for in terms of increasing the mud weight at the proper depth, it may lead to uncontrolled fluid flow to the surface. This phenomenon is typical of sediments that compact, such as sands, shales, and chalks.

Within the overpressured formation, the rock is weak because its effective stress is low. Drilling rates of penetration, therefore, increase relative to the overlying rocks. However, the mud weight needs to be carefully monitored so that it is high enough to hold back the pore fluids, but low enough not to fracture the rock. Poisson's ratio decreases with increasing pore pressure. This decrease can be predicted ahead of drilling from computed seismic attributes.

Overpressured zones may be shallow or deep. Shallow geopressure zones typically consist of gas-charged sand bodies that derive their charge from underlying gas pools which has leaked upward through a series of fractures or faults. Deep overpressure zones occur in thick shale layers that have been buried so rapidly that escape of their contained water is arrested. Geophysical techniques have proven useful in predicting abnormally pressured zone (Aminzadeh et al., 2002a).

8.6   SHALLOW GAS AND SHALLOW OVERPRESSURE

On seismic data, shallow gas-charged sand bodies are typically characterized by "bright spots," or strong reflection amplitudes. Within the overpressured sand, a very low acoustic impedance results from the lowered density and velocity due to the introduction of gas. This creates a large acoustic imped-ance contrast between the sand layer and its enclosing rocks, which yields a large reflection coefficient. They are typically encountered off the mouths of actively depositing deltas, although they also exist in stable platforms, such as the Persian Gulf, where shallow sand bodies overlie multiple hydrocarbon-bearing reservoirs.

When there are a number of these sand bodies, stacked over a deeper oil or gas pool, their overall effect is to reduce the amount of seismic energy trans-mitted to deeper reflectors. As a result, a so-called "gas chimney" forms in the seismic section. This manifests itself as a cylindrical zone of low-amplitude, disorganized reflections, which pretty much obliterates meaningful seismic imaging of any horizon (Fig. 8.4). The presence of such a chimney presages a drilling risk.

To better image gas chimneys, the "chimney cube" concept has been intro-duced that highlights vertical anomalies on the seismic data, associated with gas clouds and gas chimneys. They are used to address drilling hazards caused by shallow gas pockets and platform stability problems due to subsea mud volcanoes. Aminzadeh et al. (2002b) and Aminzadeh et al. (2013). They are also very useful for exploration of hydrocarbon targets both in highgrading

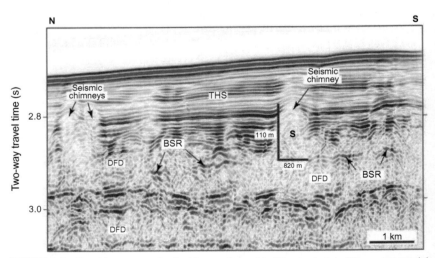

FIGURE 8.4   Gas chimneys or seismic chimneys imaged in 3D seismic data. These are potential drilling hazard locations. *Courtesy: EGEBS*. (For color version of this figure, the reader is referred to the online version of this chapter.)

prospects and better understanding of the petroleum system. Practically, chimney cubes can reveal where hydrocarbons originated, how they migrated into a prospect, and how they spilled or leaked from this prospect and created shallow gas, mud volcanoes or pockmarks at the sea bottom. Current applications of the *ChimneyCube* include detecting shallow gas and geo-hazards, distinguishing between charged and non-charged prospects, determining vertical migration of gas, and unraveling a basin's migration history. New applications of the chimney cube data include identifying potential for overpressure, Fig. 8.5 shows an example of use of gas chimneys to detect shallow gas pockets that potentially create drilling hazard.

In some cases, the pressure of the gas exceeds the mechanical strength of the overlying, uncompacted shales that serve as "seal," or the sealing capacity of faults that cut the formation. The gas may then breach the overlying sediments and can erupt at the surface as "mud volcanoes." In other cases, gas from deep pools may actually bypass shallow sand bodies and flow along a fault all the way to the surface if such fault extends up to the water bottom.

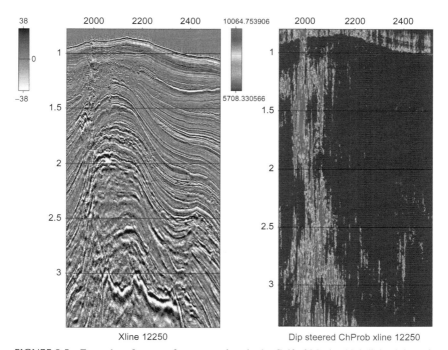

FIGURE 8.5   Examples of near-surface gas pockets in the Gulf of Mexico high-lighted through gas chimney processing (right), with the original seismic section (left). © *Society of Exploration Geophysicists, Tulsa.* (For color version of this figure, the reader is referred to the online version of this chapter.)

8.7 DEEP GEOPRESSURE

Many drilling projects including deep geopressured shales manifest themselves seismically not in amplitude, but in velocity variation. As shown in
Chapter 3, the acoustic velocity of a rock is directly proportional to its rigidity
and inversely proportional to its compressibility. Geopressured shale is essentially buried, uncompacted sediment. Thus, its rigidity is low and its compressibility is high. As a result, its acoustic velocity is low compared to the
normally compacted rocks above it. This inversion in seismic velocity forms
the basis for predicting the onset of deep geopressure.

At a well, the variation of acoustic velocity with depth may be obtained
from a velocity checkshot or vertical seismic profile (VSP) survey, sonic
log of the formations, or derived from the seismic stacking or migration
velocity analyzed from the pre-stack seismic data. All these three data sets
measure the same earth and they yield the same information after they have
been properly calibrated with each other. While the sonic log provides very
fine sampling (0.5-ft spacing) of the velocity structure, it must be corrected
to the checkshot or VSP data, which average velocity every 50 ft typically.
The seismic stacking or migration velocity is actually an imaging parameter,
not a true velocity, so its conversion to formation velocity must be determined
by calibrating it to the VSP velocity profile. At points away from the well,
where the sole source of velocity information comes from the seismic data,
the same conversion parameters are applied to the seismic velocity to derive
the predicted velocity profile of the subsurface.

In fact, there are several methods by which the onset of geopressure may
be inferred from seismic velocities:

1. From seismic stacking velocities while processing;
2. From interval velocities computed between seismic horizons or continuous
   reflections along which stacking velocities have been measured, sometimes called horizon-based stacking velocity analysis;
3. From migration velocities obtained from pre-stack depth migration; and
4. Interval velocities obtained from tomographic analysis of all possible seismic rays traversing the subsurface.

It is beyond the scope of this book to describe each of these methods in detail,
but it suffices to state that all of them yield a velocity picture of the subsurface
from which zones of lowered velocity may be interpreted. These methods differ
in the accuracy with which the desired velocity is obtained, with the accuracy
generally increasing from raw stacking velocities to tomographic analysis.

Of greater importance than the method by which these velocities are obtained
is how seismic-derived velocity is converted to pore pressure. The relationship
between pressure and velocity may be derived from pressure measurements and
sonic log at a well (Fig. 8.5). Where the range of the expected pressures to be
encountered is not fully represented at a well, theoretical modeling may be

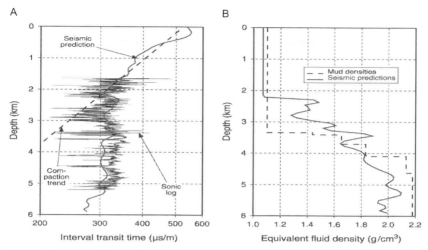

FIGURE 8.6   Part A Seismic derived transit time. B Seismic predicted geopressure. Prediction of overpressure from seismic velocities calibrated with sonic logs at wells. The velocities decrease as overpressure zones are approached. *From Kan and Swan (2001). © Society of Exploration Geophysicists, Tulsa.* (For color version of this figure, the reader is referred to the online version of this chapter.)

performed to obtain velocities corresponding to pressures outside the range actually observed at the well. Once this relationship is confirmed at other wells, it may be used to transform the velocity data to predicted pressure. Figure 8.6 shows the process of calibration of pressure at a well with sonic log and seismic-derived velocity at the same location. The seismic-derived velocities over the area provide a prediction of the over pressure prior to drilling. If a 3D velocity picture has been derived from the 3D seismic data, then likewise a 3D distribution of predicted pressure becomes available for visualization prior to drilling.

8.8   CONSTRUCTION HAZARDS

Prior to actually penetrating the earth, a drilling program starts by constructing a surface drilling platform. On land, this involves building a road to a well site, preparing the ground for a drill rig, housing facilities, material depot, and possible pipeline routes. In the deserts of the Middle East, this preparation includes finding a source of water. At sea, it involves evaluating the location of a platform or anchor points, assuming a drilling ship is not used. As such, the ground surface or the water bottom must be thoroughly investigated for hazards in erecting drilling-related structures. These hazards are referred to here as construction hazards because they are adjunct to, and not a direct part of, the actual drilling. As discussed earlier, chimney cubes derived from seismic data can be used to detect potential sea floor stability problems.

8.9   HAZARDS IN MARINE ENVIRONMENT

In the ocean, surficial hazards also occur on the water bottom. Mud volcanoes, sediment slumps, shallow faults, active channels, and soft sediments are not suitably competent to support significant loads. A detailed picture of the bathymetry, sediment distribution, and other features of the water bottom must be carefully compiled before any man-made structure is placed on top of it. High-resolution seismic surveys using sparker sources or an array of small-volume air guns are used for imaging in shallow water. Side-scan sonar recordings along the same traverse lines provide a lateral view of the water bottom, albeit along a narrow swath about the ship's traverse, to complement the vertical section obtain in the seismic profiles. If these traverses are made at close enough spacing so as not to miss any meaningful obstacles, then a fairly accurate quasi-3D view of the water bottom and the near-surface sediments can be obtained. These so-called "hazard surveys" are not only necessary for safe marine construction but also mandated by responsible federal agencies of many countries. Sunken ships, underwater pipelines, and other underwater structures obviously need to be avoided, just like transmission lines and roads on land. These obstacles must be carefully detected and mapped.

The situation is much more fortunate in deepwater (700 ft or greater depths) than in shallow water, particularly if a 3D seismic survey has already been acquired. The same data may be used to map the water bottom in detail, although with a slightly reduced resolution, in tens of feet rather than feet. All the seismic assessment tools for reservoir characterization—amplitude and attribute analysis, coherency, impedance, and others—may be brought to bear in determining the structures and sediment distribution on the water bottom. Furthermore, unlike sparker surveys, the 3D data is properly migrated so that bothersome diffractions are collapsed and the reflections are placed at their true locations in space. The data may also be visualized in its various attributes and at various perspectives—vertical and horizontal cuts, combinations of cuts, or arbitrary traverses—as the seismic volume itself.

By calibrating the measured seismic attributes of the water bottom to the composition and mechanical properties of samples taken from sediment cores, borings, or engineering tests, a detailed classification of the sediments may be constructed from the 3D seismic data. The water bottom thus becomes another horizon, like the target reservoir itself, for detailed seismic analysis.

8.10   LOOKING "AHEAD" AND "AWAY" FROM THE BIT

The drilling program is planned based on the geomechanical and geological characterization of the borehole location from the surface to the total target depth. As drilling progresses and the bit penetrates past the shallow subsurface, further questions arise which require geophysical assistance. Some of these questions are:

1. How far is the bit from the next casing or coring point?
2. How far is the bit from the top of the target reservoir?
3. What is the distance of the wellbore from faults or salt domes?
4. What does the reservoir, or any formation, penetrated by the bit look like away from the borehole?

These questions have a greater urgency in exploratory or delineation wells where the subsurface structure is likely to be uncertain and poorly understood due to paucity of well control. Seismic data attributes provide information in locations between wells.

8.11   HORIZONTAL DRILLING AND GEOSTEERING

Drilling of horizontal multilateral wells improves the productivity and injectivity index, reduces the number of wells for field development, and also optimizes the sweep efficiency for a given drainage plan in the reservoir. Horizontal drilling involves real-time monitoring of the well during the drilling phase. The well path is updated by proactive geosteering, which allows us to adjust the borehole position on the fly, to reach the geological targets, as drilling progresses. These adjustments are made based on geological information gathered while drilling. In conventional deviated drilling, the well path is steered according to a predetermined geometric path defined by the well plan and also drilled with conventional steering assemblies.

When the geological markers are poorly defined and the target tolerances are tight or the geology is complicated with offsetting faults, as in North Sea and Persian Gulf, conventional deviated drilling becomes impractical. Geosteering also integrates biostratigraphy, measurement while drilling (MWD), logging while drilling (LWD), and seismic while drilling (SWD) with the convention 3D seismic data. MWD/LWD/SWD will be discussed later. The development of deepwater fields in West Africa, Campos basin in Brazil, stacked thin layer sand-shale sequence with numerous faults in the Persian Gulf owe their success to evolving geosteering technologies. The drilling risk, as well as, cost have been reduced with the implementation of geosteering in many of these extended reach wells.

Drilling of horizontal wells comprise of the following phases:

1. Wellbore planning to define the geometry of the well path. This is based on detailed characterization of the subsurface structural and stratigraphy in 3D including geomechanical characteristics of the well path. Seismic data attributes are combined with well log data from nearby wells are used for the modeling and planning (Fig. 8.7).
2. Well path building while drilling to optimize the location of landing of the well bore.

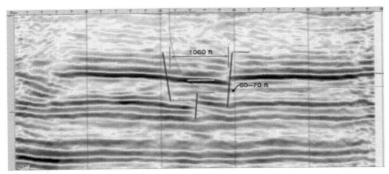

FIGURE 8.7   Planning of high angle deviated and horizontal wells begin with interpretation of seismic data. The location of faulting is important to ensure the bit stays in zone. *Courtesy: Columbine Logging (http://columbinelogging.com)*. (For color version of this figure, the reader is referred to the online version of this chapter.)

3. Navigating to ensure landing the drilling assembly in the reservoir zone.
4. Geosteering the drill bit to adjust the well bore being drilled. This is performed continuously guided by the 3D characterization model and combined with real-time logging measurement while drilling (LWD), for example, gamma ray, resistivity, sonic, nuclear, and SWD.

(Fig. 8.8) shows one application of horizontal drilling to track the reservoir boundary in a hydrocarbon reservoir. A radar system is used to allow for MWD guidance and navigation. The corresponding seismic section is used to steer the directional drilling. Well planning of high angle or horizontal drilling requires a detailed model of the structural and stratigraphic framework. The model is defined using geophysical interpretation and available geomechanical data and geological information. Drilling engineering planning is also dependent on the subsurface model. The model is used for casing design, planned deviation angle for well trajectory, formulation of drilling fluids, potential drilling hazard zones, and navigating the well trajectory. The well trajectory path to the target layer is defined and guided proactively in real time using LWD data as the well is drilled, this ensures navigating the drill bit to optimized landing in the reservoir zone and to continue with the horizontal trajectory in the pay zone as the drilling proceeds.

The implementation of geosteering as described earlier, is done using updates from the real-time MWD, LWD data that allow us to look ahead and look around the drill bit. The real-time characterization and 3D visualization provide capability for proactive geosteering. This eliminates the drilling of side tracks, reduces well stability issues, fewer doglegs, higher reservoir contact, and higher cumulative production.

Figure 8.9a shows geosteering operations where the initial geological models are updated by integration of real-time LWD data and the 3D visualization. This allows for proactive decisions in steering the drill bit through the

FIGURE 8.8   Planned horizontal well path from the initial mode, next to a conventional vertical well for coal bed methane application, from National Energy Technology Laboratory (NETL): *http://www.netl.doe.gov/technologies/oil-gas/Petroleum/projects/EP/AdvDrilling/15477Stolar.htm* (For color version of this figure, the reader is referred to the online version of this chapter.)

reservoir targets. In combination with surface seismic data, MWD and LWD tools are often used to look ahead and away from the bit. The reservoir entry and exit points are defined in the well planning stage and therefore are anticipated long before they occur. The direction of approach to the reservoir zone is determined from azimuthal resistivity measurements and changes to the well trajectory. The LWD resistivity images are used during geosteering. Figure 8.9b shows an example of wellbore planning, with geosteering. Seismic 3D volume data is used in the initial wellbore planning. The LWD data is used to fine tune the horizontal well path.

Figure 8.10 shows geosteering operations center in Saudi Aramco where the initial geological models are updated by integration of real-time LWD data and the 3D visualization. This allows for proactive decisions in steering the drill bit through the reservoir targets.

8.12   GEOSTEERING IN WELLS

After landing in the reservoir interval, geosteering of the drill bit is performed along the wellbore. Geosteering is performed in real time using 3D seismic and wire-line data as guide in initial modeling. The model is constantly updated with real-time information obtained during drilling. The challenge

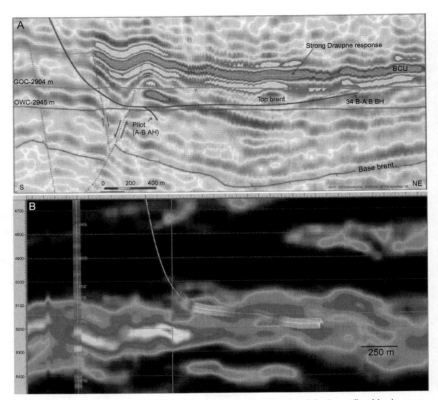

FIGURE 8.9   (a) Planned deviation path in the model is updated and further refined by incorporating seismic attributes, petrophysical data from well logs, geomechanical properties, and structural details. (b) Wellbore planning, landing of wellbore in the reservoir, and geosteering. The initial model seismic attributes like AVA is updated with LWD data in real time as drilling progresses. *(a) Courtesy of StatOil ASA. Reprinted with permission.* (For color version of this figure, the reader is referred to the online version of this chapter.)

here is to stay within the reservoir pay zone (Mutari et al., 2009). Exiting the reservoir zone during drilling would result in costly and non-productive break through. This could potentially bypass oil production along the well bore in the reservoir and affect the ultimate recovery. Geosteering allows us to optimize the well path from updates of the model with real-time measurements while drilling. The technology provides the means of steering the drill bit with reference to geological markers. The markers often are the top and bottom of the pay zone, frequently defined via gamma ray or resistivity data.

LWD and drilling event analysis information are used in real time for critical drilling decision making. Geosteering helps to maximize the drilling efficiency, optimize wellbore placement in the reservoir zone and hydrocarbon production. Geosteering is applied in many types of reservoirs especially in thin stratified and faulted zones.

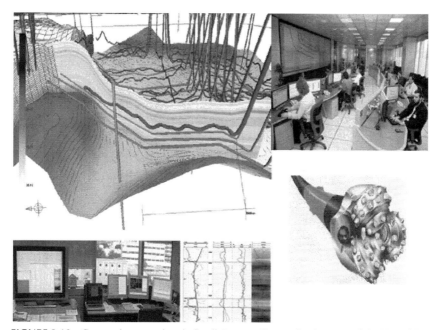

FIGURE 8.10   Geosteering operations in Saudi Aramco. Geosteering is accomplished by guiding the drill bit with real-time update of wellbore characterization using MWD and LWD data. *Courtesy: Saudi Aramco (Personal Communication).* (For color version of this figure, the reader is referred to the online version of this chapter.)

8.13   GULF GEOSTEERING CASE STUDY

Figure 8.11 shows the original depth-converted V_p/V_s volume from 3D seismic data. The right panel was depth converted with velocity model using only the available vertical wells. The velocity model is generated for input to the seismic inversion volume as part of the low-frequency model-building process. Nine vertical wells within the 75-km^2 study area were used to build the initial velocity model for depth conversion. Using this velocity model for depth conversion can introduce depth errors in excess of 50 ft, which is not accurate enough when drilling horizontal wells targeting thin sands. The left panel is image of the depth volume generated using a refined velocity model incorporating the tops from all vertical and 23 existing horizontal wells (Tonellot et al., 2011).

The target sands were found to be almost 55 ft deeper using the refined velocity model and this was later confirmed through drilling. Objective of the imaging process is to provide engineers with a 3D depth picture of stringer sand distribution to optimize the drilling of horizontal wells. To generate this depth volume, it is important to convert the inversion results from time to depth as accurately as possible as these thin sand stringers are only 20–30 ft vertically. The geosteering was performed using the refined depths for V_p/V_s

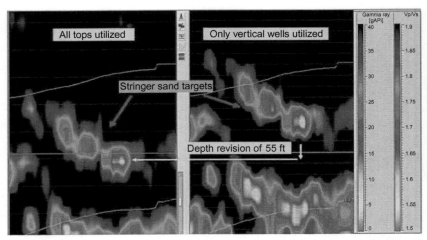

FIGURE 8.11 Original depth-converted V_p/V_s volume to the right incorporated only the available vertical wells. The image to the left is the depth volume generated using a refined velocity model incorporating the tops from the horizontal wells. The target sands were almost 55 ft deeper using the refined velocity model and this was later confirmed through drilling (Tonellot et al., 2011). *Courtesy of EAGE. Reprinted with permission.* (For color version of this figure, the reader is referred to the online version of this chapter.)

volume and penetrated the porous sand reservoir. The new model reduced the depth errors. Without using the refined velocity, the well completion in the sand would likely have missed part of the pay zone.

8.14 SWD/MWD/LWD TOOLS

In combination with surface seismic data, SWD, MWD, and LWD tools are often used to look ahead and away from the bit. In many instances, it is desirable to update the predrill estimates of formation parameters with information obtained while the well is being drilled. This is typically accomplished using MWD and LWD logs which utilize specialized drill collars and data telemetry systems that allow wire-line measurements to be made, as the well is being drilled. Because MWD/LWD systems commonly use mud-pulse telemetry, they are real-time measurements, compared to wire-line measurements, which are made only at casing points. They were developed for use in high-risk wells and for high angle deviated or horizontal wells, which were difficult to log with wire-line methods.

8.15 SEISMIC WHILE DRILLING

Rotating tricone roller cone drill bit generates strong seismic energy that propagates vertically up and down the borehole. Unlike many other seismic sources, drilling does not generate an impulsive seismic source. Instead, it continuously generates a stream of seismic energy from the drill bit. This train of "shots" propagates continuously to the surface receivers such that the recorded signal becomes a superposition of responses (Dasgupta, 2005) to the various "shots." Seismic wave propagation is reversible. When shots and receivers are interchanged, the same travel times and energies are recorded as before. Thus, instead of placing a seismic source at the surface and acoustic detectors inside the borehole, seismic energy may be generated inside the borehole and recorded by an array of receivers at the surface. Figure 8.12 shows a set up that can be considered as one type of SWD.

Borehole seismic sources that are available today have not been powerful enough to transmit sufficient energy to the surface without damaging the borehole. Since this process effectively constitutes a seismic experiment, it is called "seismic while drilling."

The force exerted on the formation by the drill bit consists of the weight of the drill string and the torque is exerted by the rotation of the bit. The forward propulsion of the bit generates the maximum compressional seismic energy along the borehole direction; only a small amount of energy propagates perpendicularly away from the borehole. Hence, for a vertical well, the cone of seismic energy that is generated propagates vertically. This is ideal for a receiver array placed around the surface location of a vertical well, but not

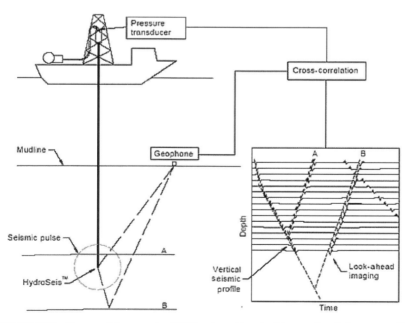

FIGURE 8.12   One possible geometry SWD, using either the bit as a downhole source. (For color version of this figure, the reader is referred to the online version of this chapter.)

for receivers placed along a separate borehole at the same depth as the bit in the drilling well. Experiments in a well test facility in Italy have demonstrated, however, that another kind of seismic energy, this one a vertically polarized shear wave, or S_v wave, is generated by the drill bit in a horizontal direction, perpendicular to a vertical borehole.

Since the recorded seismic signal at the surface consists of overlapping responses to the continuous series of drill bit "shots," they need to be separated. The individual impulses sent to the surface by the drill bit may be separated by recording the vibrations of the drill string which impart the energy to the bit and correlating it with the received signal, in a manner similar to Vibroseis correlation. An accelerometer attached to the top drive of the drill string accomplishes this task. Its recording is sent to the seismic recording truck for continuous monitoring. SWD does not disrupt the drilling process. It also provides real-time data, delayed only by the time required to process the signal. Thus, prediction ahead of the bit is made continuously.

There are several important limitations to using the drill bit as a downhole seismic source. First, the only types of bits that generate strong enough signals are "roller cone" bits that break rock through compressive failure. PDC bits, which break rock through scraping or shear failure do not generate observable energy at the surface. Second, the technique does not work well in highly deviated or horizontal wells because the axis of p-wave radiation is along the axis of the drill string and in a horizontal well, p-waves will be radiated horizontally and will never reach the surface. See Rector and Hardage (1992) for a detailed discussion of the radiation pattern of seismic waves by a working roller cone drill bit. There are other factors that affect the drill bit signal such as weight on bit (WOB) and RPM. In many instances, particularly when drilling with mud motors, the WOB may not be sufficient to generate signals at the surface. In general, more than 10,000 lbs of static weight is required to generate adequate signals at the surface.

8.16   REAL-TIME MONITORING OF DRILLING PROCESS

Successful drilling in deep, high temperature and high pressure, hostile environments is a challenging and costly endeavor with far reaching implications including economic, safety, and environmental impacts. Modeling and simulation of the drilling process combined with real-time monitoring, data acquisition, data mining, and integration are key elements of a proactive drilling strategy. Developing such an integrated strategy requires a highly instrumented and monitored infrastructure with specialized tools to detect and report problems, in addition to expert supervision to analyze and resolve issues.

Real-time monitoring of the drilling process goes back to the late 1980s. Figure 8.13 from Neill et al. (1993) shows how one can monitor the drilling trajectory using the noise generated by the drilling process as a seismic

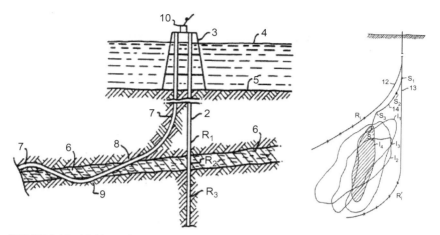

FIGURE 8.13   Multistage imaging of the subsurface earth structure.

source. The figure on the left shows the well trajectory, attempting to maximize the penetration through the reservoir layer with receivers (R_1, R_2, R_3) in another nearby well, or at the surface. The figure on the right also shows the well trajectory with the seismic measurements are made in real time repeatedly (S_1, S_2, S_3, S_4,. . .) when the drill bit is at different points at the well trajectory. The corresponding seismic images created with the drill bit at different locations are denoted by I_1, I_2, I_3, I_4. These images are updated based on the new data collected (SWD data). This information can possibly combined with MWD, logging while drilling (LWD), and coring, while drilling creating useful information for different horizontal drilling projects.

Predictive modeling and simulation of the drilling processes together with real-time assimilation of data from a variety of sensing instruments create a mirror of the drilling process. This provides critical information in real time on key drilling parameters such as well path, pressure and temperature profiles, stress, and friction conditions along the drill string and wellbore, cuttings transport conditions, well instability tendencies, pore pressure ahead of drill bit, optimal ROP, all in real time. Real-time geomechanical modeling combined with conventional and SWD data, allows efficient model updating. This is particularly important in increasingly complex environments, where the pressures, stresses, and rock properties are uncertain and maintaining a stable wellbore is extremely challenging. For example, see Kolnes et al. (2007).

8.17   LOOKING AHEAD AND AROUND A HORIZONTAL WELL

For real-time drilling geosteering application, several azimuthal LWD tools are used for imaging ahead of the drill bit and also around the bit. Most

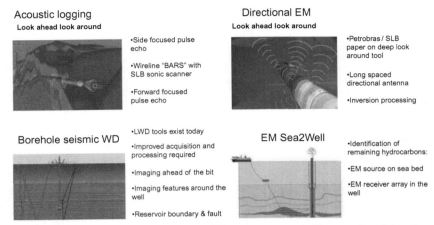

Acoustic logging
Look ahead look around

•Side focused pulse echo

•Wireline "BARS" with SLB sonic scanner

•Forward focused pulse echo

•LWD tools exist today

Borehole seismic WD

•Improved acquisition and processing required

•Imaging ahead of the bit

•Imaging features around the well

•Reservoir boundary & fault

Directional EM
Look ahead look around

•Petrobras / SLB paper on deep look around tool

•Long spaced directional antenna

•Inversion processing

EM Sea2Well

•Identification of remaining hydrocarbons:

•EM source on sea bed

•EM receiver array in the well

FIGURE 8.14 Look ahead and look around LWD measurement tools. The acoustic tools have less penetration and more limited capability than electromagnetic (EM) tools. *Courtesy of StatOil ASA. Reprinted with permission.* (For color version of this figure, the reader is referred to the online version of this chapter.)

of these tools are acoustic or electromagnetic (EM)-based azimuthal or directional tools and are used in combination with other LWD tools like sonic, gamma ray, on the borehole assembly. Figure 8.14 shows a number of measurements while drilling tools.

8.18   LOOKING ABOVE AND BELOW A HORIZONTAL WELL

Reflections from both above and below a horizontal well may be recorded if the seismic source is located within the horizontal borehole. Since borehole acoustic sources are essentially weak piezoelectric transducers, the target reflectors must lie close to the borehole (less than 50 ft at most). Furthermore, the reflectors generated energy interfere with energy that propagate principally within the borehole, the same events recorded in a normal sonic logging run—such as compressional, shear, and Stoneley (or borehole fluid) waves. Thus, presurvey planning must carefully model and evaluate the feasibility of successfully imaging the desired reflectors.

Consider a hypothetical reflection survey setup inside a horizontal borehole. For each shot, several seismic waves are excited. The strongest of these events travel within the borehole. In fact, they propagate past the farthest receiver, get reflected from the end of the hole (TD), and re-recorded in the receiver array on their return trip. The reflections from above or below the receiver are much weaker than the borehole modes. Measurements of the comparative energies of these arrival show that the Stoneley wave can be as much as 50 times stronger than the desired reflection signal.

In the plot of travel time versus offset, these reflections display hyperbolic travel time curves (or move out) which intersect the borehole arrivals at certain shot-receiver offsets. However, as the plot shows, there are regions between the borehole modes, called "quiet zones," where recording of the reflected signal will be relatively free of interference. One such zone is at very short offsets, others lie between the P and S arrivals, and between the S and Stoneley arrivals. Hence, the separation between the source and receiver array must be determined carefully from modeling so that expected reflected arrivals occur within these "quiet zones."

Typically, the roof of the reservoir comprises the top reflector, while the fluid contact (GWC or OWC) or the base of the reservoir forms the bottom reflector. In between, internal reflectors representing vertical variations within the reservoir, due to either lithological facies change or porosity variations may exist. Usually, however, the top of the reservoir and fluid contacts constitute the strongest reflectors around the borehole since these present the greatest impedance contrast with their surroundings. In fairly thin reservoirs, about less than 50-ft thick, both reflectors may generally be imaged.

One borehole seismic imaging tool is called borehole acoustic reflection system. This tool is essentially a modified sonic logging tool that consists of a 4–12 kHz acoustic transducer source and an array of 8 receiver stations spaced 6 in. apart. Each receiver station contains 4 hydrophones which are recorded separately. Hence, a total of 32 waveforms are recorded for each shot. The closest shot-receiver spacing can be varied between 30 and 50 ft, depending on where modeling indicates the "quiet zones" for borehole seismic imaging lie. Spacers placed between the source and the receivers enable this variability to be achieved. In a horizontal borehole, the entire logging tool has to be conveyed down by the hole by a drill string pipe. When deployed in the borehole, the tool constitutes a microscale seismic reflection survey, whose data can be processed like surface seismic data but with a much higher frequency content.

REFERENCES

Al-Mutari, B., Jumah, S., Al-Ajmi, H., Saleh, K.M., Burman, J.R., Reeves, D., 2009. Geosteering for maximum contact in thin-layer well placement, SPE-120551. Presented at the SPE Middle East Oil and Gas Show and Conference (MEOS).

Aminzadeh, F., Chilingar, G.V., Robertson, J.O., 2002a. Seismic methods of pressure prediction. In: Chilingar, G.V., Serebryaskov, V.A., Robertson, J.O. (Eds.), Origin and Prediction of Abnormal Formation Pressures. Developments in Petroleum Science Series, vol. 50. Elsevier, Netherlands, pp. 169–190.

Aminzadeh, F., Connolly, D., Heggland, R., Meldahl, P., de Groot, P., 2002b. Geohazard detection and other applications of chimney cubes. Leading Edge 21 (7), 681–685.

Aminzadeh, F., Berge, T., Connolly, D., 2013. Hydrocarbon Seepage: From Source to Surface. SEG Publications.

Dasgupta, S., 2005. Use of drill bit energy for tomographic modeling of near surface layers, US Patent 6868037 March 15, 2005.

Kan, T.K., Swan, H.W., 2001. Geophysics. Society of Exploration Geophysicists 66 (6), 1937–1946.

Kolnes, Ø, Kluge, R., Halsey, G.W., Bjørkevoll, K.S., Rommetveit, R., 2007. From sensors to models to visualization—handling the complex data flow. In: SPE Digital Energy Conference, Houston, Texas, 11–12 April, SPE 106916.

Neill, W.M., Aminzadeh, F., Sarem, A.M.S., Quintana, J.M., 1993. Guided Oscillatory Well Path Drilling by Seismic Imaging, Patent # 5242025.

Rector, J.W., Hardage, B.A., 1992. Radiation pattern and seismic waves generated by a working roller-cone drill bit. Geophysics 57 (10), 1319–1333.

Tonellot, T.L., Fitzmaurice, J., Khater, S., Rahati, S., 2011. Borehole Geophysics Workshop—Emphasis on 3D VSP: 3D VSP (Poster 2). EAGE, Istanbul Turkey.

Chapter 9

Geophysics for Unconventional Resources

Chapter Outline

9.1 INTRODUCTION

Unconventional resources refer to a recent trend that has been very successful in the production of gas and oil from source rocks with extremely low permeabilities. These formations are now considered as unconventional gas and oil reservoirs. The exploitation of these resources applies unconventional innovative techniques. This has added large reserves mostly in North American oil provinces. In many other parts of the world, unconventional gas and oil resources have been discovered but have not yet been assessed. In future, this new technology that is being developed in the United States will be applied

Developments in Petroleum Science, Vol. 60. http://dx.doi.org/10.1016/B978-0-444-50662-7.00009-3

worldwide to augment gas and oil production from unconventional reservoirs and contribute to needed energy supplies.

As conventional oil and gas reservoirs are depleted, unconventional gas and oil reservoirs will be developed in those basins. Conventional gas and oil resources are relatively easily to produce; unconventional resources are more difficult to develop and more costly to produce. Oil and gas in conventional sandstone and limestone reservoirs typically flow through pore spaces and sometimes through natural fractures in the rock. In tight versions of these rocks, however, the amount of pore space, the size of the pores, and/or the extent to which the pores interconnect are significantly less than in conventional reservoirs and production of oil and gas is more difficult. Major gas and oil production from these low-permeable unconventional reservoirs is facilitated by massive hydraulic fracturing treatments that increase permeability and help to reactivate natural fractures (Agarwal et al., 1979).

Horizontal drilling and multistage hydraulic fracturing technologies are applied to produce from these very low permeability gas and oil reservoirs. In multistage hydraulic fracturing, various segments of the well are isolated with packers and hydraulic fracture treatment is applied in one stage at a time (Fig. 9.1). This has the effect of creating one long section of fractured reservoir along each frac stage. Fracture length is the overriding factor for increased productivity and recovery, the objective in fracking is to maximize the length of the induced fracture. An understanding of the fracture geometry and orientation is critical in planning the development of the reservoir—in determining well spacing and field development strategies for optimizing hydrocarbon recovery.

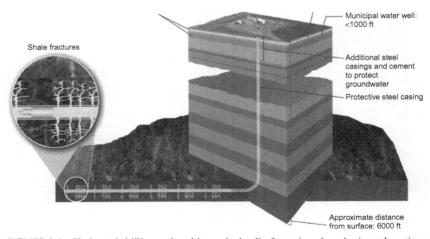

FIGURE 9.1   Horizontal drilling and multistage hydraulic fracturing along horizontal sections in unconventional resource stimulate the rocks to produce gas and oil. (For color version of this figure, the reader is referred to the online version of this chapter.)

Over the past decade, application of this drilling and completion technology has transformed the natural gas industry in North America. At present, over 60% of the natural gas produced in the United States is from unconventional resources. Production from these tight formations has changed the outlook for natural gas supply from a projected future shortage to a position of sufficient abundance. Now there is consideration for exporting excess gas volumes as liquefied natural gas.

Tight gas development began with the Barnett shale in Texas (Fig. 9.2), Bakken play in North Dakota, and Montana in the United States. Other shale gas and oil resources are in Haynesville, Marcellus, Cotton Valley, Eagle Ford, Fayetteville, Woodford, Niobrara, Horn River, and Utica formations. Higher oil prices and recent technological advances have provided economic incentives and a driver for exploiting these oil-bearing formations that historically had been very difficult to attain commercial production. The application of horizontal drilling and multistage hydraulic fracturing to tight oil formations (Fig. 9.3) has been applied to produce shale gas and has given new life to these previously low-producing or unproductive oil reservoirs, in many areas. This has reversed the declining trends in oil production in many producing fields.

The classification of unconventional resources as a reservoir is mainly technology and economics driven. Natural faulting and fracturing are critical factors controlling the present-day stress distribution, which in turn influences hydraulically induced fracture system development. Stimulation ultimately enhances reservoir drainage, yielding economically viable hydrocarbon production (Rutledge and Phillips, 2003).

Holditch (2006) points out that the resource-triangle concept (Fig. 9.4) or logarithmic-normal distribution is valid for all natural resources in all basins in

FIGURE 9.2   Core samples from Barnett shale reservoir. Extremely low porosity and permeability. (For color version of this figure, the reader is referred to the online version of this chapter.)

FIGURE 9.3 Microphotograph of impermeable pores in tight gas formation. *Source: http://energy.USGS.gov.* (For color version of this figure, the reader is referred to the online version of this chapter.)

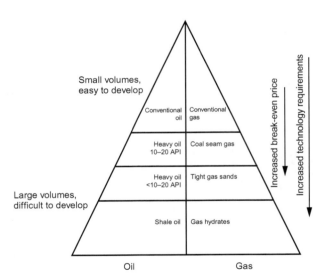

FIGURE 9.4 Resource triangle for hydrocarbons showing conventional smaller volumes that are relatively easy to develop at the apex with larger resources toward the base of the triangle that are in unconventional reservoirs and more difficult to develop and produce. *After Holditch, 2006.* (For color version of this figure, the reader is referred to the online version of this chapter.)

the world, so it is logical to believe that large volumes of gas and oil in unconventional reservoirs will be found, developed, and produced in every basin that now produces significant volumes of oil and gas from conventional reservoirs. While the conventional proved reserves are about 1.5 trillion barrels of crude oil as

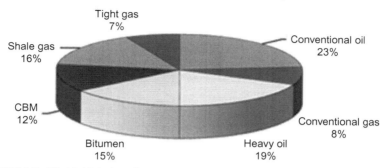

FIGURE 9.5   Worldwide hydrocarbon resources. Conventional resources make up less than a third of the total resource; the rest is from various unconventional resources. *Courtesy of CGG.* (For color version of this figure, the reader is referred to the online version of this chapter.)

estimated by USGS, the world's estimated oil reserves from unconventional resources such as heavy oils and tar sands alone are about three trillion barrels. Although the unconventional resource base is large, its production cost is more and unconventional resources are economical only when oil prices are high. The technologies deployed in exploiting these resources are new and some still evolving (Fig. 9.5).

9.2   GEOPHYSICAL DATA CONTRIBUTION TO UNCONVENTIONAL RESERVOIRS

Shales and tight sands in unconventional reservoirs, until recently, were considered to be source rocks or reservoir seals (Figs. 9.2 and 9.3). Due to their low permeability, fluids cannot be produced without stimulation of the reservoir with induced fractures or hydraulic fracturing. The matrix permeabilities for tight gas and oil shales are in the order of 100s of nanodarcies. Because of such low permeabilities, the distance that gas or oil molecules can diffuse through the reservoir rocks during production is limited. For an optimal hydraulic fracturing program, geomechanical parameters such as brittleness and closure pressure are important. In shale gas and oil production, the extent of both the natural fracture system and the frac-induced fracture network need to be determined. Although every unconventional play is unique, the key properties governing production potential generally include total organic content (TOC), brittleness, natural fractures, and closure stress (Ouenes, 2012).

von Lunen et al. (2012) define the challenges and opportunities for geophysical input for unconventional resources. They point out that "instead of the traditional concerns with trap mapping, spill points, and degree of fill, unconventional resource plays require information on reservoir quality, fracability, fracture networks, and the stimulated rock volume (SRV) resulting from frac-completion programs." The extent of the SRV is based on rock mechanical properties, the initial state of stress both vertically and laterally, and the effectiveness of seals and barriers to isolate the producing rock media. For unconventional resources in a continuous system, the presence of hydrocarbon is already known, so the reservoir fill is not the issue. Geophysical data are used for defining the boundaries of the reservoir unit like in conventional reservoirs.

Geophysical data in unconventional plays provide information for characterizing and mapping of the fracture system. This reduces the risk in the selection of drilling location. Most of the unconventional gas and oil formations have been found onshore. Acquisition techniques of land seismic data are therefore undergoing transformations to improve imaging of the subsurface in drilling and development of these unconventional plays. An important aspect of drilling for any petroleum is predetermining the success rate of the operation. Seismic data are acquired, analyzed, and interpreted for determining the drilling locations. These seismic surveys can define the best areas to drill for tight gas reserves. Inversion of seismic data provides information on facies distribution, mineral content, and rock strength. These provide the preferential drilling locations. Seismic-derived elastic properties and Ant Tracking software for fault-fracture swarms in 3D seismic data together with recorded microseismic data can provide an understanding to the relationship between reservoir quality and production.

9.3   ROCK PROPERTIES FOR UNCONVENTIONAL RESOURCE DEVELOPMENT

Hooke's law describes the relationship between strain and stress. Stress and strain are functions of the elastic properties of rocks and represent the fundamentals of hydraulic fracturing. The deformation and fracturing are caused by stressing the rock with hydraulic pressure in the borehole. The stress induced during hydraulic fracturing causes sufficient strain on the formation leading to rock failure.

The geomechanical properties are estimated between existing wells based on Young's modulus (E) and Poisson's ratio (v) derived from inversion of seismic data volumes. Estimation of these rock properties is imperative for drilling and well completion by hydraulic fracturing. Estimates of the stress state of rocks before drilling are useful for predicting areas at risk for wellbore failure. These properties, therefore, have direct bearing on the placement of wells, reservoir productivity, and the safety issues in fracturing completion strategy. It is assumed that the subsurface rocks *in situ* are constrained horizontally, that is, the horizontal strain is zero in their natural state, and that the rocks are undergoing elastic deformation.

Goodway et al. (1997) introduced amplitude versus offset (AVO) inversion techniques to derive Lamé's parameters (λ: lambda, μ: mu) and density

(ρ: rho) from prestack seismic data. Mu is the shear modulus. V_p is the velocity of compressional or P-waves and V_s, of shear of S-waves.

$$\mu = \rho V_S^2 \qquad (9.1)$$

$$\lambda = V_P^2 \rho - 2\mu \qquad (9.2)$$

These elastic moduli can be transformed to estimate Young's modulus, Poisson's ratio, bulk modulus, and shear modulus. These moduli are important in estimating how rocks will fracture and whether the fractures will remain open. Variations in Young's modulus and Poisson's ratio should be expected due to variations in lithology, porosity, fluid content, and cementation in reservoir rocks. K, E, and σ are the elastic moduli needed.

$$K = \lambda + \frac{2\mu}{3} \qquad (9.3)$$

$$E = \frac{9K\lambda}{3K + \mu} \qquad (9.4)$$

$$\sigma = \frac{\lambda}{2(\lambda + \mu)} \qquad (9.5)$$

Optimal drilling locations for unconventional reservoirs are usually where naturally occurring fracture networks are already present. Fractures exist naturally in most reservoir rocks. They need to be abundant and the fracture network connected enough to be conduits for hydrocarbons to flow in producing wells. Often despite the abundance of fractures, they are usually difficult to detect and quantify. Characterization and modeling of reservoir fractures are achieved by integrating geophysical, geological, and engineering data. The hydraulic fracturing process usually enhances these fractures. If the wells are drilled directly into the best areas to develop the reserves, costs of development will be minimized. Interpretation of seismic data can assist the drilling engineers to locate these areas and determine where and to what extent drilling directions should be deviated.

To predict production results and to design optimum drilling and hydraulic fracturing programs, data analysis of various seismic attributes and inversion of 3D seismic volume, sometimes in conjunction with microearthquake data (Maity, 2013), is performed. The interpretation of these seismic attributes provides estimates of geomechanical properties that are calibrated with well logs and core measurements. From the rock strength and stress regime, reservoir and drilling engineers can evaluate the ductility or brittleness of target rocks.

Brittle rocks are more susceptible to fracking (like Barnett shale). The rock strengths—Young's modulus, Poisson's ratio, unconfined compressional strength—and reservoir rock stresses—horizontal and vertical stresses, closure stress, fracture initiation pressure—are required. Young's modulus, Poisson's ratio, and brittleness can be derived from analysis of seismic attributes. From these parameters for strength of rocks and whether the rock will fracture sufficiently under stress, reservoir engineers can evaluate the ductile and brittle behavior of the rocks. The principal stresses and the directions of the horizontal stresses are deduced and used for the well drilling and completion planning.

9.4 SEISMIC INVERSION FOR ROCK PROPERTIES

Analysis of seismic attributes and seismic data inversion provides estimates of the dynamic rock strength parameters; these must be calibrated with geomechanical properties derived from well logs and core measurements. Understanding what parameters can be extracted from seismic data and determining the data reliability are key to interpreting seismic attributes. Seismic attributes are correlated with production data in unconventional resource plays through empirical data correlations. Well data from cores and logs measure the static elastic parameters that are applied in well design. In older and harder rocks, the dynamic and static rock strengths are similar. For younger, less consolidated rocks, the calibration with well data usually results in application of a scale factor to equalize them. From the estimation of stress regime, we deduce how the induced fractures from hydro-frac treatment will propagate. This information is critical for planning well orientations and for designing completion programs.

Seismic attributes such as lambda, mu, and rho (LMR) provide Lamé's modulus, shear modulus, and density. The LMR attributes also identify areas of high TOC. In some unconventional resource play areas, multi-component seismic data calibrated with well logs are being used in the inversion process. This provides information that is used in well planning for optimizing well paths and for selecting hydraulic fracture locations. Gray et al. (2010) describe a method for the estimation of rock strength and the three principal stresses between wells from 3D seismic data. This technique has been applied successfully for White Second Speckled Shale Formation in Colorado shale gas plays in Alberta. The results show that both rock strength and stress can vary considerably over small distances on the order of 0.1 km. Reservoir lithology and local stress regime define the best locations for successful hydraulic fracturing. The best hydraulic fracture networks are produced where the rock is brittle and there is little differential stress.

Seismic attributes, along with well logs and core measurements, can provide these parameters. Estimation of existing fractures and whether they will open and stay open is the key information needed for well completion. Rock strength is estimated from AVO analysis. LMR attributes as described earlier. The stress in the rock is estimated from azimuthal AVO, azimuthal velocities, and multicomponent seismic fracture analysis. Azimuthal anisotropy in seismic amplitudes is due to differential horizontal stresses. Image logs, cores, and downhole measurements are used to calibrate seismic azimuthal anisotropy of elastic properties of rocks.

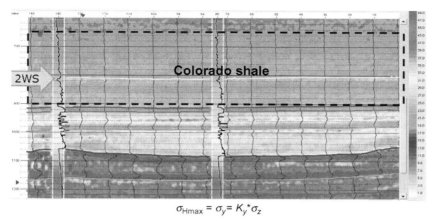

$$\sigma_{Hmax} = \sigma_y = K_y{}^* \sigma_z$$

FIGURE 9.6   Maximum horizontal stress from seismic attributes calibrated at the wells. *From Gray et al. (2012). Courtesy of CGG.* (For color version of this figure, the reader is referred to the online version of this chapter.)

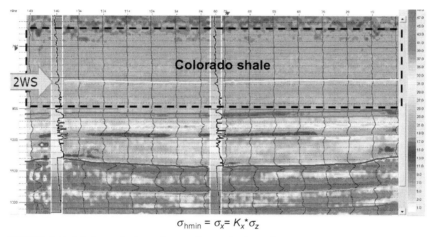

$$\sigma_{hmin} = \sigma_x = K_x{}^* \sigma_z$$

FIGURE 9.7   Minimum horizontal stress from seismic attributes calibrated at the wells. *From Gray et al. (2012). Courtesy of CGG.* (For color version of this figure, the reader is referred to the online version of this chapter.)

The stress state of the reservoir defines how the induced fractures from hydraulic fracture treatment will propagate in the rocks. This provides the maximum and minimum stress and the existence of microfractures in the rocks and whether the fractures will remain open (Figs. 9.6 and 9.7). This information is used by engineers to decide on the horizontal well orientations and in designing completion programs. Usually image logs, core measurements, drilling information, and other downhole data are used to calibrate seismic azimuthal anisotropy effects in amplitude and velocity.

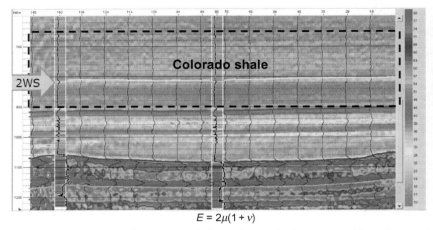

$$E = 2\mu(1 + \nu)$$

FIGURE 9.8 Seismic attribute computed from 3D seismic data can provide estimates of Young's modulus (*E*) which is related to the brittleness of rocks. Higher *E* signifies areas with more brittle rocks that are prone to frac with relative ease. *From Gray et al. (2012). Courtesy of CGG.* (For color version of this figure, the reader is referred to the online version of this chapter.)

Young's modulus can be computed from the seismic attributes (Fig. 9.8) and provides a measure of brittleness of the rocks (assuming density is constant). Differential horizontal stress ratio (DHSR) is another important parameter for prediction of hydraulic fractures. The direction of the DHSR (Fig. 9.9) indicates the estimated direction of maximum horizontal stress σ_{Hmax}.

As described earlier, in the unconventional low-permeability reservoirs, production of fluids can be established by fracking or stimulating the reservoir with fractures. In the planning of an optimal hydraulic fracturing program, geomechanical factors such as brittleness and closure pressure are important. These can be estimated between existing wells based from Young's modulus and Poisson's ratio derived from inversion of seismic data. Areas with higher Young's modulus (*E*) and lower Poisson's ratio (*v*) are brittle and more fracable. An estimate of fracability could be defined by (*E/v*); larger values of this ratio relate to areas with higher fracability. This information can be used by drilling engineers in designing the completion program in wells.

For interpretation of seismic attributes and understanding of what information can be extracted from the attributes, model templates are created using well measurements from cores and well logs. The models provide a range of rock physics parameters showing how seismic attribute changes can be used in predicting the rock and fluid properties. Variations of seismic attributes calibrated with the rock physics models are used to predict changes in mineralogy, porosity, pore shape, fluid inclusions, stress, etc.

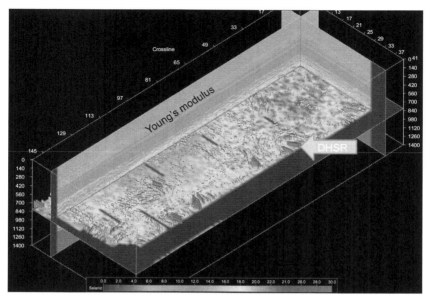

FIGURE 9.9   Young's modulus estimated from seismic attributes in 3D seismic. The arrow indicates plates for the differential horizontal stress ratio (DHSR). DHSR $= \sigma_{Hmax} - \sigma_{Hmin}/\sigma_{Hmax}$. The size of the plate is proportional to the magnitude of DHSR and the direction of the plate shows the direction of the local maximum horizontal stress. *From Gray et al. (2012). Courtesy of CGG.* (For color version of this figure, the reader is referred to the online version of this chapter.)

The interpretation of attributes and prediction is constrained by petrophysical parameters from well logs, core data, and the geologic understanding. In LMR crossplots of MuRho–LamdaRho, lines of constant E are approximately lines of $\mu\rho$, assuming constant density and thus yielding an estimate of E. The brittle-to-ductile transition is a function of Young's modulus (Goodway et al., 2012). In LMR crossplots, higher $\mu\rho$ (Mu–Rho) values can be interpreted as increasing brittleness (Fig. 9.10). There is a relationship between higher EUR and lower $\lambda\rho$ (LamdaRho), which suggest that this attribute is suitable for mapping potential production performance in gas shales (Fig. 9.11).

For subsurface modeling, seismic 3D data volume is acquired using wide azimuth surveys that are especially designed for the unconventional plays. These surveys are performed with full-offset range and a full set of azimuths. Wide-azimuth seismic surveys enhance the total value of these shale assets through improved imaging quality, resolution, and more accurate reservoir characterization. The acquired data are processed using reservoir-specific seismic processing workflow. The reservoir characterization of the processed volumes provides prediction of the sweet spots and geomechanical and reservoir models.

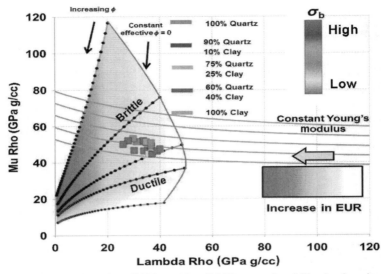

FIGURE 9.10   Model template of LMR crossplot of MuRho versus LamdaRho showing seismically derived attributes and corresponding estimated ultimate recovery (EUR). *From Goodway et al. (2012). Courtesy of GeoScienceWorld.* (For color version of this figure, the reader is referred to the online version of this chapter.)

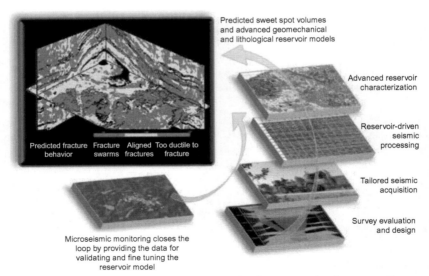

FIGURE 9.11   Integrated geophysical reservoir solution for unconventional resources. *Courtesy of CGG.* (For color version of this figure, the reader is referred to the online version of this chapter.)

9.5 AMPLITUDE VARIATION WITH ANGLE AND aZIMUTH

As was discussed in Chapter 3, AVO is used in many areas, primarily for detecting gas-saturated reservoirs. A variation of AVO called Amplitude Variation with Angle and aZimuth (AVAZ) is used for estimation of geomechanical properties and fracture characterization. Significant changes in AVAZ due to the presence of fractures are observed in full azimuth 3D seismic data. Estimated production over the field area can be derived by combining lithological, geomechanical, and stress properties using seismic data calibrated with well measurements. The lithological and geomechanical properties that would provide the optimum match with hydrocarbon production need to be derived for each survey using multiattribute correlation and will vary from one shale play location to another.

The AVAZ effect is due to local changes in the direction and intensity of the azimuthal anisotropy of both the rocks' rigidity and its seismic velocity, usually caused by changes in the direction and intensity of fracturing and/or stress. We can measure the AVAZ effects in the acquired seismic data and analyze them to create an estimate of the crack density and orientation of the fracture trend. Figure 9.12 shows an example where AVAZ attribute shows the relative fracture density and strike in the reservoir.

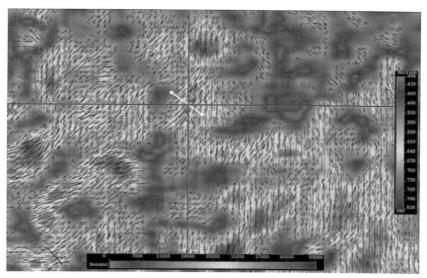

FIGURE 9.12   Image of processed AVAZ intensity and strike indicating the relative fracture density and strike at this location in the reservoir. The well path marked in white crosses this level at the cross hairs indicating moderately intense fracturing with a strike of NNE at the well location. *Courtesy of CGG.* (For color version of this figure, the reader is referred to the online version of this chapter.)

The AVAZ measurements come from seismic amplitude data and have similar resolution, while velocity variations with azimuth (VVAZ) measurements have much lower resolution. Reservoir caprocks are generally not fractured and so do not exhibit this azimuthal anisotropy in rigidity. Different rocks behave differently under the same stress loads, for example, sandstone tends to fracture, while shales tend to flex under similar stress loads. So, AVAZ can be an indicator of the presence of both fractured reservoirs and unbreached caprock.

Success rates of over 80% have been achieved for mapping fracture trends in unconventional gas plays using seismic azimuthal anisotropy measurements (Goodway et al., 2012). This significantly impacts drilling success in many areas. These measurements also show the fracture strike and so, by identifying where the gas is coming from, they can be used to avoid drilling into depleted pools. New methods of calculating the stress state can now identify the best areas for fracking to ensure maximum productivity from every well.

9.6   SEISMIC ANISOTROPY FOR FRACTURE DETECTION

Unconventional gas plays in shale rely on the presence of natural fractures to enhance or create permeability in the reservoir and for planning stimulation by fracking. Presence of fractures causes significant changes in 3D seismic data. These changes appear as variations in seismic amplitudes and velocities with azimuth and constitute azimuthal anisotropy. Because of S-wave anisotropy, S-waves split into two waves (birefringence), a fast and a slow mode. The split S-waves are very sensitive to fractures and can provide information about fracture density (fracture porosity) and orientation (directions of preferred permeability).

Seismic azimuthal anisotropy can be used to pinpoint higher producing areas of natural fractures in fractured unconventional gas reservoirs such as tight gas, gas shales, and coal bed methane. These measurements also show the fracture strike and so, by identifying where the gas is coming from, they can be used to avoid drilling into depleted pools. Results from a carbonate reservoir in offshore Qatar indicates amizuthal amplitude anisotropy that coincide with conductive fractures, outlined in red with high gas production rate. Low amplitude anisotropy area, outlined in green is an area of low fracture intensity. See Fig. 9.13.

9.7   MICROSEISMIC AND HYDRAULIC FRACTURE MONITORING

Microseismic events that are caused by subsurface fluid injection and production (SFIP) in both conventional and unconventional fields could be monitored by a specially designed seismic network (Dasgupta et al., 2008). Likewise, the induced seismicity generated by hydraulic fracturing and the SFIP operation can not only provide information to improve safety but also obtain additional information

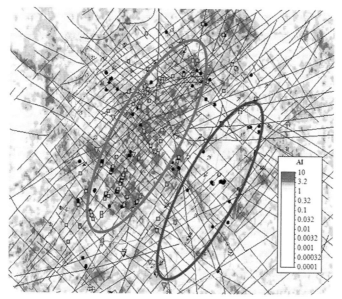

FIGURE 9.13 Normalized azimuthal amplitude anisotropy intensity map with overlay of interpreted faults (black). Mud-loss points from wells (yellow) indicate permeable fracture zones. Outlined in red is the known conductive fracture corridor which correlates with high anisotropy intensity values, and outlined in green is an area of low fracture intensity. *Courtesy of CGG.* (For color version of this figure, the reader is referred to the online version of this chapter.)

about the reservoir. Figure 9.14 shows a setup for monitoring microseismic events. Traditional microseismic mapping determines the location and magnitude of the event. When microseismicity is observed over time, operators may observe patterns of seismicity related to SFIP activities. Real-time microseismic monitoring and analysis could improve the effectiveness and safety of fracking which is crucial in certain situations.

Multistage hydraulic fracturing, where the shale is fractured under high pressures at several stages located along the horizontal section of the well, is used to create conduits through which gas can flow. The tight gas plays in shale formations in North America have been successfully applying hydro-fracture stimulations and monitoring the fracture geometries using microseismic data. The frac stimulation locations are selected in areas with the most brittle reservoir rocks usually estimated from seismic interpretation. The microseismic technique has been applied for the assessment of reservoir deliverability and the recovery factor (Mayhofer et al., 2008).

Optimizing unconventional gas recoveries requires far more wells than would be the case in conventional natural gas operations. The horizontal wells are drilled with multiple horizontal sections up to 2 km in length and are

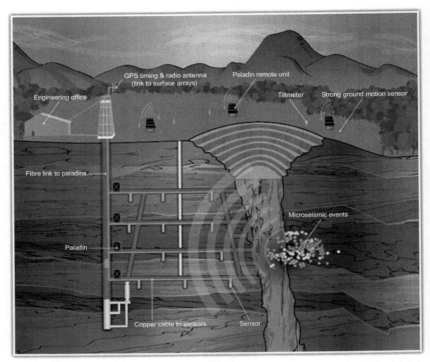

FIGURE 9.14 A typical microseismic monitoring system with both surface and subsurface sensors. With color-coded microseismic events associated with different stages of fracking. *Courtesy of ESG Solutions Inc., www.esgsolutions.com/CMImages/Monitoring%20Applications/Frac/Frac-schematic. png.* (For color version of this figure, the reader is referred to the online version of this chapter.)

fractured in multiple stages to drain the reservoir to the maximum extent possible. Microseismic imaging allows oil and gas companies to visualize where this fracture growth is occurring in the reservoir. However, as a technology-driven play, the rate of development of shale gas may become limited by the availability of required resources, such as fresh water and fracture proppant, or by availability of drilling rigs capable of drilling horizontal wells several kilometers in length.

Figure 9.15 depicts the evolution process of the microseismic events in a reservoir with hydraulic fracturing treatment. The frac treatment increases stress in the reservoir rock with the influx of pressure as the frac fluids are pumped in. This decreases the inherent stability in the existing cracks and weak zones along natural fracture zones, along bedding planes of deposition and other heterogeneities. New fractures are also created along the present-day principal stress direction. The resulting rock failure and slippage generate microtremors similar to earthquakes. These triggered microseismic events are detected using sensitive seismic sensors or geophones. The geophones are

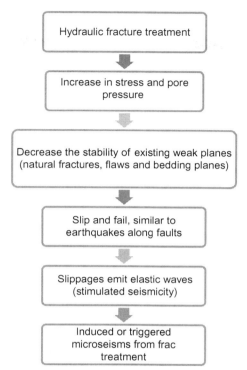

FIGURE 9.15 Hydraulic fracture treatment in a tight reservoir, evolution of microseisms, and their monitoring with microseismic technique. (For color version of this figure, the reader is referred to the online version of this chapter.)

deployed in a monitor well or sometimes buried below the ground surface. From the recording of these detected microseismic events, the fracture treatment results can be mapped.

Due to pore pressure changes during hydraulic treatment, the elastic reservoir rock matrix undergoes a stress failure known as Coulomb failure criteria where the rock cracks open. Slip occurs along the existing fractures, and faults in the reservoir rocks or new fractures are formed. The stress failure in rocks generates small earthquakes or microseisms (Toda et al., 2005). Detecting the direct arrivals from these microseisms that are triggered during the fracturing process provides monitoring of hydraulic fracture treatments. The length of the generated fractures and the geometry of the induced fracture system may be determined using microseismic data. In a naturally fractured reservoir, hydraulic treatments reactivate natural fractures and locally enhance permeability. Figure 9.16 shows the generation of the microseismic events from hydraulic fracturing treatment in tight gas reservoirs and their monitoring. Monitoring of microseismic events provides useful information about the treatment results and the level of success of the rock fracturing from the treatment.

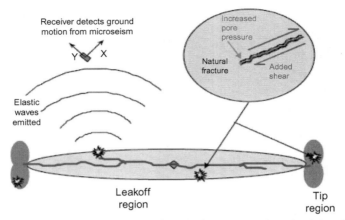

FIGURE 9.16 Evolution and detection of microseismic events (or microseisms) from hydraulic fracturing. The receivers in an observation well or buried below earth surface detect ground motion from the microseisms. (For color version of this figure, the reader is referred to the online version of this chapter.)

Passive monitoring of microseismic emissions is applied for estimating the stimulated rock volume (SRV), future production, and recovery factor in unconventional tight shale and sand reservoirs that are fracked. An effective way to detect the fracture orientation and fracture density in the subsurface reservoir rocks is by using birefringence or splitting of microseismic shear wave events. As shear waves hit the fractured medium, they split into two components which have fast and slow arrivals on the seismogram. The polarization angle (θ) of fast shear wave component is parallel to the fracture and indicates the fracture orientation. The time delays (δt) observed between the slow and fast shear waves provide an indication of fracture density (Maity, 2013). Normalized time difference, divided by total travel time (or the length of the ray path), is proportional to fracture density along the seismic ray path (Vlahovic et al., 2003).

9.8   MICROSEISMIC MONITORING CASE STUDY

The case study of a successful monitoring of a multistage hydraulic frac treatment from Cardium tight sandstone reservoir located in Alberta, Canada, is described here. The microseismic survey design for monitoring this treatment is shown in Fig. 9.17; the monitor well is drilled 120 m from the horizontal path of the treatment well. For microseismic monitoring, seismic three-component sensors or geophones were conveyed in the vertical observation or monitoring well. The geophones were placed 10 m apart and they straddled the target Cardium formation with geophones located above and below the reservoir. Vibrator trucks were used to generate a controlled

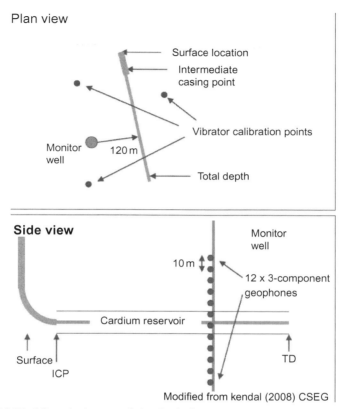

FIGURE 9.17 Microseismic survey design for the frac treatment in Cardium reservoir, Alberta. *Source: SEG-Duhault, 2012. Courtesy of SEG.* (For color version of this figure, the reader is referred to the online version of this chapter.)

seismic source at various points on the surface. This is a calibration procedure for determining the orientation of well geophones and is performed prior to frac treatment in the production well. The drilled horizontal well was treated with a gelled oil frac at several stages.

The fracture stages are approximately 100 m apart. The microseismic data are recorded continuously during each frac stage. The recorded data are processed for locating the microseismic events that are generated from the rock fracturing. The computed microseismic events from this experiment were clustered along the frac stages of the horizontal wellbore and trending NE–SW with a clear separation between the frac-wings and event "clouds" (Fig. 9.18) for the different frac stages. The frac stages with higher closure stress exhibited lower number of microseismic events. The events recorded in this experiment are weak and are probably controlled by existing faults or are due to changes in lithology.

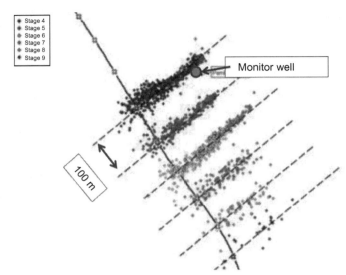

FIGURE 9.18 Cardium formation hydro-frac treatment showing in plan view location of micro-seismic event hypocenters clustered along the frac stages of the horizontal borehole. Frac wing is asymmetric to the NE up to 240 m. *Source: SEG-Duhault, 2012. Courtesy of SEG.* (For color version of this figure, the reader is referred to the online version of this chapter.)

The monitoring well is located on the NE side of the treatment well. The majority of these event clouds trended approximately N45E–S45W from the treatment well. In addition to the consistent frac-wing NE azimuth, it was noted that the frac-wings were not symmetric; they were asymmetric with more events in the NE than in the SW. No matter which side the observation well was located relative to the treatment well (based on observations in other monitoring experiments in the area), a larger portion of the microseismic events were found trending to the NE (Duhault, 2012).

The frac width and length, frac height, and frac azimuth from the principal stress direction can also be estimated from the microseismic data. These observations are then compared against the time-synchronized frac-pumping curves. Passive seismic is the only technique that can directly measure the creation of fracture drainage in the stimulated reservoir. The data provide the reservoir engineer a measure of SRV and an estimate of fracture complexity. This information is also used in reservoir simulation model (Maxwell et al., 2012).

The microseismic events were not all contained within the Cardium reservoir zone. As seen from the side view in Fig. 9.19, the events were recorded 100–150 m above and 30–50 below the tight Cardium zone. This indicates that the frac treatment had propagated away from the treated zone and created fractures in the zones above and below the zone. This could be due to preexisting faults or fractures and changes in the rock fabric to more fracable rocks.

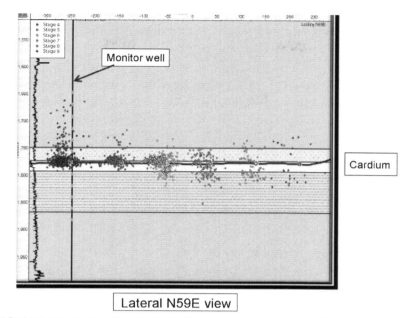

FIGURE 9.19   Cardium formation hydro-frac treatment showing from side view locations of microseismic event hypocenters clustered along the horizontal borehole. Some events are scattered up to 150 m above the treatment zone, also events 50 m below this zone. *Source: SEG-Duhault, 2012. Courtesy of SEG.* (For color version of this figure, the reader is referred to the online version of this chapter.)

9.9   INTEGRATION OF GEOPHYSICAL TOOLS FOR UNCONVENTIONAL RESOURCES DEVELOPMENT

Monitoring of hydraulic fracture treatments from unconventional reservoirs using microseismic data has become imperative for measuring induced permeability that facilitates production from tight formations. Microseismic events result from an interaction between the stress induced upon the reservoir rock formation during hydraulic frac stimulation, the stress regime within the rock, and existing fractures in the rock. Along with other tools, microseismic data are used to estimate the effective fracture volume or the SRV. Microseismic data provide an additional source of calibration and validation of fractures that are predicted from seismic attributes. Mapping of microseismic events allows us to define the connectivity between reservoir zones, to identify induced fracture orientation, to estimate stress tensor magnitude in rock formations, and to estimate the orientation of the stress field. Microseismic analysis also provides characterization of seismic anisotropy to calibrate seismic geomechanical models through seismic numerical models and to establish their relationship to rock fluids and geomechanical properties.

9.10   CALCULATING STIMULATED RESERVOIR VOLUME BASED ON DEFORMATION

The main goal of hydraulic fracturing is to stimulate the tight shale and also sandstone reservoirs. To assess the effectiveness of the process, stimulated reservoir volume (SRV) is calculated. SRV is defined as the volume of a reservoir which is effectively stimulated to increase the well performance. Aside from fracking, a similar concept is used to assess the effectiveness of steam flooding or other EOR operations to create new or activate existing fracture networks. In low-permeability reservoirs, production is increased by the creation of fracture networks; therefore, it is useful to use SRV to describe the efficacy of the stimulation treatment.

Most estimates of SRV, albeit not too accurate, are limited to drawing boxes around microseismic event maps and adding up the 3D volume where the events are observed. This is because not all microseismic events are guaranteed to increase production in a reservoir, while other events act to deform the reservoir more than others. SRV can also be calculated from seismic deformation which is the change in shape or size of an object due to an applied force. Thus, seismic deformation is a measure of the deformation of a rock mass as a result of tensile, compressive, or shear forces exerted on the rock by operations such as hydraulic fracture treatments. It describes the density of seismic moment release and is proportional to the degree of fracturing which has occurred. For example, volumes that have small seismic deformation will tend to not be extensively fractured, whereas volumes that have large seismic deformation will either have a complex network of many small fractures or a number of large fractures, or both. In terms of hydraulic fractures, volumes with high seismic deformation will show increased permeability and, therefore, would likely contribute to reservoir production more effectively.

Additional spatial and temporal analysis of seismic deformation can be used to identify regions with higher deformation (accounting for >90% of the observed deformation), and therefore increased permeability, resulting in an increased contribution to production. Figure 9.20 shows SRV calculated from seismic deformation information.

9.11   INTEGRATED APPROACH FOR FRACTURE ZONE CHARACTERIZATION

Conventional seismic data have been extensively used as a tool to understand the subsurface. Ouenes et al. (2004) showed how both prestack and poststack 3D seismic data can be used for fracture zone characterization when used with engineering and geologic data.

Next, we calculate and study various seismic attributes including log-derived properties mapped within the volume (using ANN property prediction workflows). Selected seismic attributes can be applied in combination in order

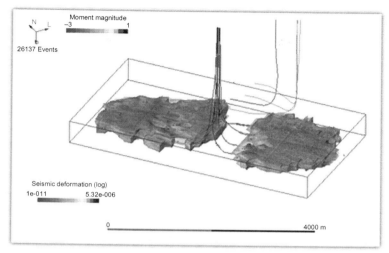

FIGURE 9.20   Example SRV calculated for a multiwell multistage fracture operation, courtesy of ESG Solutions.

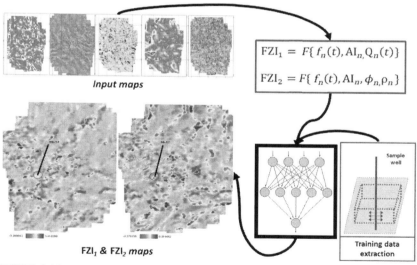

FIGURE 9.21   Different aspects of fracture zone identification. *Source: SEG-Duhault, 2012. Courtesy of SEG.* (For color version of this figure, the reader is referred to the online version of this chapter.)

to create a new set of attributes that would be considered as fracture zone identifiers (FZI).

The expressions shown in Fig. 9.21 are neural network-based nonlinear mapping of different attributes such as acoustic impedance, absorption, frequency, similarity, and porosity. There are two FZI attributes. FZI_1 takes purely seismic-derived attributes, while FZI_2 makes use of some log-derived properties such as porosity and density as well. Once the attribute volumes have been calculated, training datasets are extracted based on *a priori* information. In this case, known fracture dominated permeability zones within the perforated intervals of wells as shown in the figure at the bottom right hand side were used for the training model. Other examples could include core or image log data. The datasets are trained to generate FZI property volumes as shown. The zones of interest were further tested using available well logs.

9.12 QUANTIFYING THE RISKS ASSOCIATED WITH INDUCED SEISMICITY

As shale oil and gas production expands, applications of hydraulic fracturing become more widespread. The concerns regarding the potential risks associated with hydraulic fracturing also grows. Much of these concerns are around the chemicals used in the process and the potential for water contamination, however, there are also concerns about triggered seismicity emanated along with induced microseismicity from the fracking process. The process under which the induced seismicity is created is shown in Fig. 9.15. The public concern has prompted major studies, for example, the National Academy of National Science (2012), to better understand the risk factors associated with induced seismicity from hydraulic fracturing or other SFIP operations.

New research is being conducted to distinguish between induced seismicity from frac treatment and triggered seismicity or small earthquakes. Investigations are in progress to quantify and mitigate the risk associated with creating man-made earthquakes. Among those are Zobak (2012) and Maxwell et al. (2012).

REFERENCES

Agarwal, R.G., Carter, R.D., Pollock, C.B., 1979. Evaluation and performance prediction of low-permeability gas wells stimulated by massive hydraulic fracturing. JPT 31 (3), 362–372.

Dasgupta, S.N., Jervis, M., 2008. Passion for passive seismic in reservoir management. Saudi Aramco Journal of Technology, Spring 2008, 79–86.

Duhault, J.J., 2012. Cardium Formation Hydraulic Frac Microseismic: Observations and Conclusions. SEG Technical Program Expanded Abstract.

Goodway, W., Chen, T., Downton, J., 1997. Improved AVO fluid detection and lithology discrimination using Lamé petrophysical parameters: "Lambda-Rho", "Mu-Rho", and "Lambda/Mu fluid stack", from P and S inversions. In: 67th Annual Internat. Mtg., Soc. Expl. Geophys., Expanded Abstracts, pp. 183–186.

Goodway, B., Monk, D., Perez, M., Purdue, G., Anderson, P., Iverson, A., Vera, V., Cho, D., 2012. Combined microseismic and 4D to calibrate and confirm surface 3D azimuthal AVO/LMR prediction. Leading Edge 31 (12), 1502–1511.

Gray, D., Delbecq, F., Schmidt, D., 2010. Estimating in situ, anisotropic, principle stresses from 3D seismic. In: EAGE Annual Technical Conference, Barcelona.

Holditch, S.A., 2006. Tight Gas Sands, Distinguish Author Series, Journal Petroleum Technology/ SPE, June 2006.

Maity, D., 2013, Integrated Reservoir Characterization for Unconventional Reservoirs using Seismic, Microseismic and Well Log Data, Ph.D. Dissertation, University of Southern California.

Maxwell, S., Raymer, D., Williams, M., Primiero, P., 2012. Tracking microseismic signals from the reservoir to surface. Leading Edge 31 (11), 1300–1308.

Mayhofer, M.J., Lolon, E.P., Warpinski, N.R., Cipolla, C.L., Walser, D., Rightmire, C.M., 2008. "What is Stimulated Reservoir Volume?" Paper SPE 119890, Presented at the 2008 SPE Shale Gas Production Conference, Fort Worth, TX, 16–18 November.

National Academy of National Science, 2012. Induced Seismicity Potential in Energy Technologies (http://dels.nas.edu/Report/Induced-Seismicity-Potential-EnergyTechnologies/13355).

Ouenes, A., February 2012. Seismically Driven Characterization of Unconventional Shale Plays CSEG Recorder.

Ouenes, A., Robinson, G., Zellou, A.M., 2004. Impact of Pre-stack Seismic on Integrated Naturally Fractured Reservoir Characterization, SPE Asia Pacific Conference on Integrated Modeling for Asset Management. Kuala Lumpur, Malayasia.

Rutledge, J.T., Phillips, W.S., 2003. Hydraulic stimulation of natural fractures as revealed by induced microearthquakes Carthage Cotton Valley gas field east Texas. Geophysics 68, 441–452, reprint.

Toda, S., Stein, R., Richards-Dinger, K., Bozkurt, S., 2005. Forecasting the evolution of seismicity in Southern California: animations built on earthquake stress transfer. J. Geophys. Res. 110, 1029–1067.

Vlahovic, G., Elkibbi, M., Rial, J., 2003. Shear-wave splitting and reservoir crack characterization: the Coso geothermal field. J. Volcanol. Geotherm. Res. 120, 123–140.

von Lunen, E., Jensen, S., Leslie-Panek, J., 2012. Strategies in geophysics for estimation of unconventional resources. Leading Edge 31 (9), 1090.

Zobak, M.D., 2012. Managing the seismic risk posed by wastewater disposal, Earth Magazine.

Index

Note: Page numbers followed by "*f*" indicate figures, and "*t*" indicate tables.